Passive Solar Architecture Pocket Reference

T0139212

This handy pocket reference contains a wealth of information on a range of topics including: the principles of passive solar building and passive house, a ten-step design and build strategy, calculating solar irradiance, factors affecting the choice of building materials, passive heating and cooling principles and techniques in different climates, the Passivhaus Standard, natural and augmented lighting and notes on technology and building occupation. The book also includes conversion factors, standards, resources and is peppered throughout with helpful illustrations, equations, explanations and links to further online resources.

Ideal for practitioners, architects, designers, consultants, planners, home builders, students and academics, and those working in development contexts, the book is intended to act as an aide memoire, a reference supplement, a resource and an overview of the field. Rich in background detail, the book also includes at-a-glance tables and diagrams, equations and key definitions.

David Thorpe is a lecturer on one planet living at the University of Wales Trinity Saint David, a consultant on renewable energy and sustainable building, and author of several academic books and numerous articles. He is also Founder/Patron of One Planet Council.

Passive Solar Architecture Pocket Reference

David Thorpe

LONDON AND NEW YORK

First edition published 2018
by Routledge
2 Park Square, Milton Park, Abingdon, Oxon, OX14 4RN

and by Routledge
711 Third Avenue, New York, NY 10017

Routledge is an imprint of the Taylor & Francis Group, an informa business

© 2018 Taylor & Francis

Library of Congress Cataloging-in-Publication Data
Names: Thorpe, Dave, 1954– author.
Title: Passive solar architecture pocket reference / David Thorpe.
Description: Abingdon, Oxon; New York, NY: Routledge, 2018. |
Includes bibliographical references and index.
Identifiers: LCCN 2017024917| ISBN 9781138501287 (hb) |
ISBN 9781138806283 (pb) | ISBN 9781315751771 (ebk)
Subjects: LCSH: Solar houses—Design and construction—Handbooks, manuals, etc. | Solar energy—Passive systems—Handbooks, manuals, etc. | Architecture and energy conservation—Handbooks, manuals, etc.
Classification: LCC TH7414 .T55 2018 | DDC 728/.370472—dc23
LC record available at https://lccn.loc.gov/2017024917

ISBN: 978-1-138-50128-7 (hbk)
ISBN: 978-1-138-80628-3 (pbk)
ISBN: 978-1-315-75177-1 (ebk)

Typeset in Goudy
by Cenveo Publisher Services

Contents

List of figures and tables

Figures

Tables

1 Basic concepts

Passive solar and passive house architecture are separate but closely related and overlapping approaches to sustainable building. This is illustrated in Figure 1.1 and explained below. This book covers both passive solar and passive house design approaches, as they have much in common, and it combines the principles of each as far as possible.

1.1 Passive solar and passive house

The goal of passive solar architecture is to optimise the use of naturally available light and heat. This is achieved by designing the

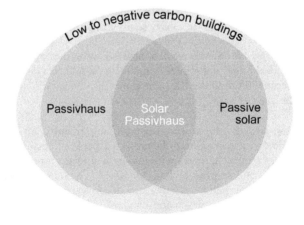

Figure 1.1 Venn diagram showing how both passive solar and Passivhaus buildings are part of the class of low to negative carbon buildings, and how passive solar buildings can also be Passivhaus, but each can be independent of the other

building with respect to the exterior conditions and location in such a way as to capture and keep thermal energy from the sun within the fabric of the building (if heating is required) or to prevent the heat gaining access in the first place (if cooling is required). A fundamental aspect of 'passive' solar design is sensitive orientation with respect to the sun for energy capture, lighting and shading.

The goal of passive house design is similar but slightly different. It is to manage thermal energy gain and loss, regardless of the source, so that a minimal amount of energy is needed to heat and cool the building. This entails controlling the number of air changes per hour and eliminating 'thermal bridges', which conduct heat or cold through the building's exterior skin. Passivhaus (passive house in English) is a strict performance-related standard for this approach.

'Passive' solar is distinct from 'active' solar, which refers to solar thermal or electric technology. In principle, 'passive' means having no operable devices – for example, motors and pumps for heating or cooling, which are considered 'active'. However, in practice most 'passive buildings' have some active components, such as blinds or heat exchangers.

Clearly, the two concepts – passive solar and passive house – overlap, but at their extremes, depending on the climate, can give rise to different building forms.

- The ideal form for a passive solar building in a temperate climate would be longer on the north–south-facing side than the east–west-facing sides, with double or triple-glazed windows on the sun-facing side and very small or no windows on the opposite side.
- The ideal form for a passive house design would be a cube. This is because it minimises the surface area to volume ratio, limiting heat loss.

Nowadays, passive solar design is approaching maturity. Since not every building can face south or be unshaded, passive house design is more widely applicable than traditional passive solar. In principle, a passive house building can be built anywhere. The same cannot be said of passive solar buildings. The signal design feature of traditional passive solar was long, sun-facing windows. These are not always needed or possible for passive house buildings.

In both cases, a comfortable interior climate is assured by using ventilation based on convection, controlled wind ingress, zoning and, where appropriate, heat exchangers or heat dumps, among other techniques.

Figure 1.2 Traditional passive solar architecture: the 'Black Forest' passive solar 'earthship' in Colorado Springs, Colorado. Made of rammed earth and reused tyres, its south-facing conservatory traps solar heat and stores it overnight in the thermally massive structure. It functions well in this location and climate, but would be inappropriate in, say, a hot dry climate or a city

Source: Michael Shealy, licensed under Creative Commons

Figure 1.3 This Passivhaus apartment block was built by construction company Rhomberg in the residential park Sandgrubenweg in Bregenz, Austria. Interestingly, two almost identical blocks were built: one to Passivhaus standard and the other to a 'low carbon' standard. They were subsequently monitored for energy use. The Passivhaus was supposed to have an annual heating requirement of 9.03 kWh per square metre and the low-energy house 33.23 kWh per square metre. In practice, the occupants in the passive house consumed 39.9 kWh per square metre and those in the low-energy house 36.3 kWh per square metre. The reason was that the occupants in the Passivhaus had the heating over two degrees higher than the designers had planned for: 22.1°C (71.78°F) instead of 20°C (68°F). This shows how occupancy behaviour can frustrate the best intentions of designers

Source: Andreas Praefcke, licensed under Creative Commons 3.0

Figure 1.4 A negative carbon house that is neither fully passive house or passive solar. The Solcer House, designed by Cardiff University's Welsh School of Architecture, combines the Welsh Low Carbon Research Institute's Solcer (smart operation for a low-energy region) programme with a research project about making buildings as power stations, conducted by Swansea University with industry partners Tata Steel, NSG Pilkington and BASF. Its foundations contain locally produced low carbon cement and its walls are of a structural insulated panel system (SIPS) made from expanded polystyrene (EPS) or extruded polystyrene foam (XPS). A 17 m^2 transpired solar collector on the south-facing wall sits below a 40 m^2 integrated photovoltaic roof pitch which is alleged to produce a surplus of electricity. Windows are high performance, there is a ventilation heat recovery system and a heat pump operating on the heated air from the transpired solar collector during cold weather. Surplus electricity is stored in lithium-ion batteries and surplus heat in a water store

Source: Author

Natural light is allowed in to the extent that it balances the need to avoid glare and the need to use artificial lighting.

The advantages of designing buildings using these principles are:

- it saves energy and running costs over the lifetime of the building;
- it provides comfort in all seasons and climates;

Figure 1.5 425 Grand Concourse Development, Mott Haven, Bronx, NY, USA (latitude 41°N). A Passivhaus certified 241 unit 100% affordable residential tower with a charter school, medical and community facilities. Joint venture project between Trinity Financial and MBD Community Housing Corp.

Source: DattnerArchitects, New York

- it is a safe investment and resilient into the future;
- there is added value every year through decreased operation costs;
- a longer useful life with high quality standard;
- it contributes to climate change protection.

The chief goal which unites these approaches is therefore to provide comfort for occupants with minimum need for additional energy, while mitigating climate change.

Experience over the last 30 years has helped to determine success-ful design in different climates throughout the world and to reduce build costs to the same or only slightly higher than conventional buildings. Lifetime operational costs are considerably cheaper – yielding, on average, savings of 85 per cent.

Success is achieved by using design tools and modelling to establish the needs and requirements of all functions in a building and their inter-relationships. Energy savings are maximised by placing spaces in the most advantageous position for daylighting, thermal control and solar integration. This process may also reveal opportunities for multi-ple functions to share space and reduce the footprint of the building.

These techniques have now been refined to such an extent that modern passive house buildings can be visually indistinguishable from conventional buildings. Of course, they can also be beautifully designed. The essential point is that there is no longer a necessity for passive solar buildings to have a particular visual style. It is all about the design.

1.2   General principles

'Passive solar' refers to the 'in use' phase of a building. But it would be self-defeating to make this phase zero carbon and ignore the other three phases: material sourcing, construction and ultimate disman-tling or demolition. Therefore it is useful to speak of 'zero carbon' or 'negative carbon' buildings. As a class, they would include passive solar and passive house. These buildings would be at least zero carbon on balance – that is, when totalling the impacts of all four phases.

'Zero carbon' does not mean refraining from using materials containing the element carbon, but avoiding the emission of greenhouse gases in these four phases. Any unavoidable emissions may be balanced by actions which remove carbon dioxide from the atmosphere and store it in the building fabric, or which generate renewable energy. A building which generates more renewable energy than it consumes over its entire lifetime is termed 'negative carbon'.[1] In summary, this entails:

- minimising the use of fossil fuel energy during the supply chain and process of construction;

- encouraging the use of materials which store atmospheric carbon in the fabric of the building, such as timber products;
- constructing and managing the building in such a way that it minimises the emission of greenhouse gases during its lifetime and eventual demolition;
- encouraging the capture, generation and even export of renewable energy.

To achieve this, the following features are needed:

- the capture of sufficient solar energy for lighting and comfort heat, with shading to avoid overheating;
- favouring the use of 'natural' and cellulose-based materials (which store atmospheric carbon in the building);
- making the structure airtight (no unwanted draughts);
- making the structure breathable (i.e. permeable to water vapour);
- making it durable, resilient, low-maintenance, fire- and weather-resistant;
- incorporating a large amount of insulation.

1.3 Assessing the location

The design of a building should be appropriate to its location. The world is divided into principal climate zones. These affect the type of

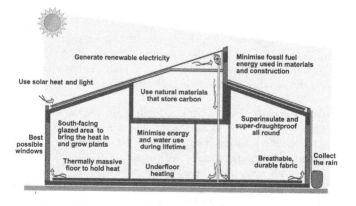

Figure 1.6 Desirable features of zero-carbon, passive solar buildings

bioclimatic or solar architecture that is appropriate. The Köppen–Geiger system is one of the most widely used climate classification systems (see Figure 1.7):

A: Tropical/megathermal climates
B: Dry (arid and semiarid) climates
C: Temperate/mesothermal climates
D: Continental/microthermal climates
E: Polar and alpine climates.

Each of these has several types and subtypes. Each particular climate type is represented by a two- to four-letter symbol.

Leslie Holdridge's Life Zone Classification system is another global bioclimatic scheme for the classification of land areas, particularly useful for humidity (see Figure 1.8).

1.3.1   Matching dwelling form and room layout to latitude and climate

For latitudes above 25° it is necessary to capture solar heat. To do this:

1.   the equator-facing glazing area should be at least 50 per cent greater than the sum of the glazing area on the east- and west-facing walls;
2.   orientation is longer on the east–west axis;
3.   this axis should be within 15 degrees of due east–west;
4.   at least 90 per cent of the sun-facing glazing should be completely shaded (by awnings, overhangs, plantings) at solar noon on the summer solstice and unshaded at noon on the winter solstice;
5.   the room plan should – if it is a dwelling – incorporate the main living rooms on the equator-facing side, with utility rooms, less-used rooms and garage, if any, on the pole-facing side;
6.   morning rooms are typically bedrooms;
7.   on the side away from the equator windows should be kept to a minimum and as small as possible for lighting to minimise heat loss;
8.   this wall should also have high thermal mass or/and be externally insulated, to retain heat in the building.

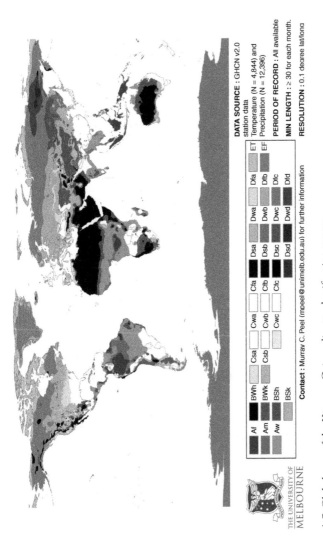

	A f	B Wh	C sa	C wa	D sa	D wa	D fa	E T
	A m	B Wk	C sb	C wb	D sb	D wb	D fb	E F
	A w	B Sh	C sc	C wc	D sc	D wc	D fc	
		B Sk			D sd	D wd	D fd	

DATA SOURCE : GHCN v2.0
station data
Temperature (N = 4,844) and
Precipitation (N = 12,396)

PERIOD OF RECORD : All available

MIN LENGTH : ≥ 30 for each month.

RESOLUTION : 0.1 degree lat/lona

THE UNIVERSITY OF
MELBOURNE

Contact : Murray C. Peel (mpeel@unimelb.edu.au) for further information

Figure 1.7 Global map of the Köppen–Geiger climate classification system

Source: M. C. Peel, B. L. Finlayson and T. A. McMahon, University of Melbourne, licensed under Creative Commons

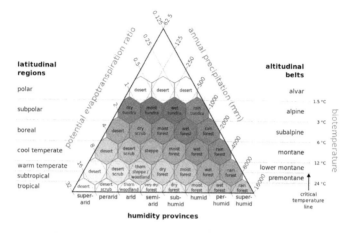

Figure 1.8 The Life Zone Classification system

Source: Creative Commons by SA

For latitudes under 25°, or where topography significantly impacts insolation, the opposite should be the case. Bedrooms, for example, need light in the morning. The whole building needs to be protected from low-angle heat.

Around 25° there is some leeway depending on local conditions. In these mid-latitudes different parts of a building may be used in the winter and summer, as equator-facing rooms become too hot and occupancy is switched in summer to rooms on the non-equator-facing side (not shown in Figure 1.9, left hand plan).

1.3.2 Spatial layout to optimise climate protection

Different layouts for building developments are appropriate according to the climate. Figure 1.10 compares two types for a hot climate which might be subject to wind, as in the Middle East and North Africa. Figure 1.11 shows how, in a housing estate in higher latitudes, privacy does not need to be compromised in order to let all houses have a south-facing aspect for solar gain and solar panels.

Table 1.1 The shape of the building has different requirements according to the local climate

Climate	Elements and requirements	Purpose
Warm, humid	Minimise building depth	for ventilation
	Minimise west-facing wall	to reduce heat gain
	Maximise south and north walls	to reduce heat gain
	Maximise surface area	for night cooling
	Maximise window wall	for ventilation
Composite	Control building depth	for thermal capacity
	Minimise west wall	to reduce heat gain
	Limited equator-facing wall	for ventilation and some winter heating
	Medium area of window wall	for controlled ventilation
Hot, dry	Minimise equator-facing and west walls	to reduce heat gain
	Minimise surface area	to reduce heat gain and loss
	Maximise building depth	to increase thermal capacity
	Minimise window wall/ window size	to control ventilation, heat gain and light
Mediterranean	Minimise west wall	to reduce heat gain in summer
	Moderate area of equator-facing wall	to allow winter heat gain
	Moderate surface area	to control heat gain
	Small to moderate window size	to reduce heat gain but allow winter light
Cool temperate	Minimise surface area	to reduce heat loss
	Moderate area of pole-facing and west walls	to receive heat gain
	Minimise roof area	to reduce heat loss
	Large window wall	for heat gain and light
Equatorial upland	Maximise north and south walls	to reduce heat gain
	Maximise west-facing walls	to reduce heat gain
	Medium building depth	to increase thermal capacity
	Minimise surface area	to reduce heat loss and gain

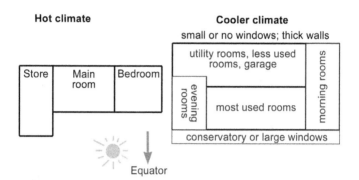

Figure 1.9 Sample optimum room layouts in dwellings for hotter (left) and cooler climates. In higher latitudes, sides facing the equator would have conservatories or large window areas to capture solar heat. Nearer the equator this is not necessary and, in fact, it may be important to keep out solar gain by use of smaller windows and shutters. Morning rooms face the rising sun. Evening rooms face the setting sun

Figure 1.10 Hot climates: inappropriate (left) and appropriate (right) spatial layouts for settlements. Organic, non-grid layouts provide shade and can be designed to block winds, preventing issues with wind funnelling. Grid layouts with random orientations borrowed from other climates, and wide spacing of buildings, do not provide shade or wind shelter. Orientation is important, however, for solar thermal and PV

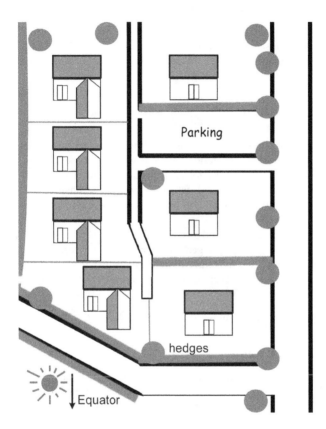

Figure 1.11 Higher latitudes: in this housing estate each property has both privacy and an equator-facing aspect and roof to maximise the potential for the use of solar energy. Grey circles are trees, grey lines are hedges (preferably) or fences

1.4   Building form

The form of a building has an effect on its energy use. It is described by reference to its surface area to volume (S/V) ratio and form factor (FF).

1.4.1   Surface to volume ratio

A low surface area to volume (S/V) ratio is better for a building that wishes to conserve energy for heating. This is the ratio between the external surface area and the internal volume. It is a measure of compactness:

$$\text{Compactness } C = \frac{\text{Volume}}{\text{Surface Area}}$$

Size is also a factor: a small building with the same form as a larger one will have a higher S/V ratio. Buildings with the same U-values (or R-values – see section 3.1 for an explanation), air-change rates and orientations, but differing S/V ratios and/or sizes may have significantly different heating demands. This has the following consequences:

- small, detached buildings should have a very compact form (square is close to the perfect optimum, the circle);
- larger buildings may have more complex geometries;
- high S/V ratios require more insulation to achieve the same U-/R-value.

In temperate zones, aim for an S/V ratio ≤ 0.7 m²/m³.

1.4.2   Form factor

The heat loss form factor (FF) is a measure of the compactness of a building in the form of a ratio of the external area of the building (not including the ground contact area, but including the roof) to the floor area. In short:

$$\text{Form Factor} = \frac{\text{Heat Loss Area}}{\text{Treated Floor Area}}$$

This ratio can be anything between 0.5 and 5. A lower number indicates a more compact, efficient building. Passivhaus buildings aim to achieve 3 or less. Once the FF is over 3, achieving the Passivhaus standard efficiently becomes more challenging.

The FF metric permits comparisons of the efficiency of the building form relative to the useful floor area. It is even more relevant than the

	Type	Form factor	Efficiency
	End mid-floor apartment	0.8	Most efficient
	Mid-terrace house	1.7	
	Semi-detached house	2.1	
	Detached house	2.5	
	Bungalow	3.0	Least efficient

Figure 1.12 Types of home and their form factors

Source: NHBC

S/V ratio because it favours buildings that require less floor-to-floor height. Most building uses do not require volume but floor area. The more compact the form, the lower the ratio. Large buildings (e.g. 172,800 ft² over 12 storeys) have a much more efficient form than small buildings or large high-bay buildings for heating load (but not cooling, where the opposite is true). Achieving a lower FF also reduces the resources required and the cost.

A building with a more complex form is also likely to have a higher proportion of thermal bridges and increased shading factors that will have an additional impact on the annual energy balance.

The effect of form on total energy consumption for a given floor area is reduced as buildings increase in size. Besides permitting greater design flexibility, this lets designers use daylighting and natural ventilation cooling strategies also to reduce energy demand, as these require one dimension of the building to be relatively narrow (between 45 and 60ft (14 and 18m)).

Good form factor   Surface area + 10%   Surface area +20%

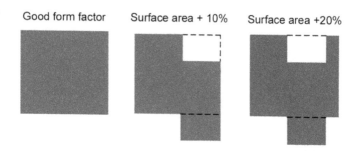

Figure 1.13 An increase in the S/V ratio of 10% (the building in the middle) would require 20 mm of insulation more than the good form on the left to achieve the same level of insulation. The one on the right (a 20% higher S/V ratio) would require an extra 40 mm of insulation

Example

A small office of 20,000 ft² (1,800 m²) having a narrow two-storey form, ideal for natural ventilation and daylighting, may have a FF ratio of 0.88, whereas a deep square plan may have one of 1.02. For the former to have the same enclosure heat loss coefficient as the latter, its overall average enclosure U-value (or R-value) would need to be 1.02/0.88 = 16 per cent lower (higher in the case of R-value). This would require a significant decrease in the opaque wall area U-value (increase in R-value), a reduction in window area or a higher-specification window.

Note

1   On average the carbon impact of the construction of conventional buildings is between 10 and 20 per cent of their use during their lifetime.

2 Calculating the solar irradiance

To design a passive solar building it is necessary to know how much solar energy is available. This is derived from the solar irradiance, which is the rate at which the radiant energy from the sun arrives on a given unit area of a surface. 'Solar gain' is the term for the amount of thermal energy that a building might derive from the solar irradiance (SI).

2.1 Definition of solar irradiance

Irradiance is described in the SI unit of watts per square metre (W/m^2). Emitted radiation may be called radiance, with the same units. The incident (arriving) solar radiation is called insolation and is expressed in terms of irradiance per time unit, such as kilowatt-hours per square metre per day. If the time unit is an hour and the energy unit a watt ($W/m^2/hr$), then the figure for insolation is the same as that for the irradiance.

Four characteristics of incident solar energy are of interest:

- the spectral content of the light;
- the radiant power density;
- the angle at which the incident solar radiation strikes a collector surface; and
- the radiant energy from the sun throughout a year or day for a particular surface.

Solar radiation at a location is usually modelled by totalling three components: direct beam shortwave radiation, sky-diffuse and ground-reflected radiation. See the *Solar Energy Pocket Reference Book* for further information.

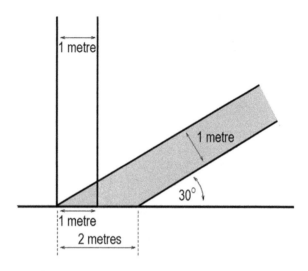

Figure 2.1 The angle of incidence affects the amount of solar energy received on a horizontal surface

Source: Author

Passive solar architecture is mainly concerned with solar wavelengths that are used for heating and daylighting – in particular, space heating (and cooling). A Campbell–Stokes sunshine recorder may be used to measure the number of hours of bright, direct sunshine.

It is also important to know the local wind patterns because of their cooling capacity. If local reports suggest wind may be a significant factor in building performance, it might be measured with a wind logger for direction, intensity and annual, seasonal or diurnal pattern.

Knowledge of sun paths is fundamental in designing building façades to admit light and passive solar gain, as well as reducing glare and overheating to the building interior. How to determine the influence of the sunshine is described in the next pages.

2.2   The declination angle

Declination (δ) is the angular distance of an object in the sky perpendicular to the celestial equator. The celestial equator is a projection from

Figure 2.2 Solar radiation may be measured with an electronic solar insolation meter

Source: Frank Jackson

a point at the centre of the earth, of its equator onto the celestial sphere. By convention, angles are positive to the north, negative to the south.

The declination angle of the sun in the sky varies throughout the year due to the tilt of the earth on its axis of rotation as it circumnavigates the sun and gives us the seasons. The axial tilt is 23.45° and the declination angle varies plus or minus up to this full amount. At the

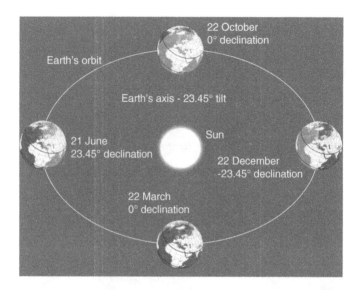

Figure 2.3 The declination of the celestial equatorial plane relative to the plane of the axis of rotation of the earth varies throughout the year due to the earth's axial tilt

Source: Author

equinoxes the value is 0°, at the 22 December solstice it is -23.45° and at the 22 June solstice it is 23.45°. It is given by the equation:

$$\delta = \sin^{-1}\left\{\sin\left(23.45°\right)\sin\left[\frac{360}{365}\left(N-81\right)\right]\right\}$$

[1]

where N is the number of the days in the year.

2.3   The elevation (or altitude) angle

The solar elevation or altitude angle (ϵ) is the angular 'height' of the sun in the sky measured from a horizontal plane which varies according to the latitude, the time of day and the time of the year, from 0° at sunrise and sunset to 90° if the sun is directly overhead.

For harvesting solar power it is usually essential to know the maximum elevation angle for each day of the year, which occurs at solar noon and is dependent upon the latitude and declination angle (δ). It is given from the following equations:

In the Northern Hemisphere:

$$\epsilon = 90 - \varphi + \delta$$

In the Southern Hemisphere:

$$\epsilon = 90 + \varphi - \delta \tag{2}$$

where φ is the latitude of the location.

2.4   The azimuth angle

The azimuth angle (α) is the compass direction from which the sun's beams arrive. The usual convention is to use a north-based system in which 0° represents north, east is presented by 90°, south by 180° and west by 270°, etc. It is found by the following equation:

$$\cos \alpha = \frac{\left(\sin \epsilon \cdot \sin \varphi - \sin \delta \right)}{\cos \epsilon \cdot \cos \varphi} \tag{3}$$

where φ and δ are negative in the Southern Hemisphere.

2.5   The equation of time

To adjust for the earth's 23.44° axial tilt and the eccentricity of its orbit we use the *equation of time*, which removes a discrepancy between the actual apparent solar time, which directly tracks the motion of the sun, and the fictitious mean solar time:

$$EoT = 9.8 \sin(2B) - 7.53 \cos(B) - 1.5\sin(B)$$

where:

$$B = \frac{360}{365.24}\left(N - 81 \right) \tag{4}$$

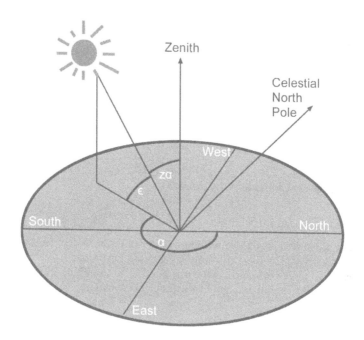

Figure 2.4 The azimuth angle (α) is measured eastward from the north. The zenith angle ($z\alpha$) is measured from the vertical. The elevation angle ($\in$) is measured from the horizon

Source: Author

in which N = the number of the day since the start of the year. (81 is the number of days since 1 January of the spring equinox.)

The Local Standard Time Meridian (LSTM) for the location is needed next. This is based on the distance in degrees of the earth's longitude from Greenwich Mean Time (GMT). One hour = 15° (360°/24 hours), so the equation for LSTM is:

$$LSTM = 15° \cdot \Delta T_{GMT}$$

[5]

where ΔT_{GMT} is the difference between local time and GMT in hours.

We then need to compensate for the variation in solar time during one time zone. This is done by applying the LTSM to the equation of

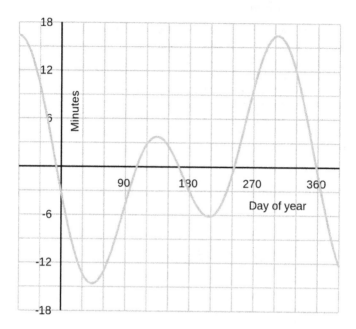

Figure 2.5 The equation of time. Local time varies compared to solar time as shown by a sundial throughout the year. When above the axis, solar time will appear fast relative to local time; when below the axis it will appear slow

Source: Wikimedia Commons, Zazou

time to reach the time correction factor (TCF). The earth rotates 1° every four minutes, so the equation is:

$$TCF = 4(Longitude - LTSM) + EoT \tag{6}$$

Local solar time is now found by adjusting the local time (LT) using the time correction factor as follows:

$$LST = LT + \frac{TCF}{60} \tag{7}$$

2.6   Sun path diagrams

Sun path diagrams are an important tool to understand how much solar gain there will be at different times of day throughout the year. They are used to evaluate the solar altitude and azimuth angles for a given latitude, longitude, time of year and time of day. Sun path diagrams represent a horizontal or vertical projection of the imaginary sky dome placed over a building site. These may graphically assess solar exposure of a reference point (e.g. a building or a part of a building) throughout a year. In a sky dome, the horizontal lines represent altitude and the vertical lines represent the azimuth. A vertical sun path diagram represents the projection of the sky dome on the vertical plane.

Caution: a sun path diagram might give impractical results in sites where a cloudy sky is dominant.

The procedure is as follows:

1. Transition the time of interest, LST, to local solar time, as above in equation [7].
2. Determine the declination angle (δ) based on time of year, as above in equation [1].
3. Read the solar altitude and azimuth angles from the appropriate sun path diagram on the link below. Diagrams are chosen based on latitude; linear interpolations are used for latitudes not covered. For values at southern latitudes change the sign of the solar declination.

Sun path diagrams for each 1° of latitude for the Northern and Southern Hemisphere are available from: http://bit.ly/1mCSJQv. Below is an illustration of how they are applied.

The chart may be superimposed upon a photographic panorama of the site to determine if and where shading is likely to occur throughout the day and year caused by nearby objects and landscape features. An alternative method is to do it by drawing. Standing in the relevant spots on the site and looking due south (north in the Southern Hemisphere and irrespective of the orientation of the array), draw a line showing the uppermost edge of any objects that are visible on the horizon (either near or far) onto the sun path diagram. This line is called the horizon line. A shading factor is calculated as follows: if the area of the diagram that falls between sunrise and sunset is termed T

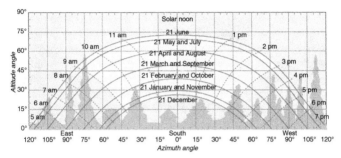

Figure 2.6 Azimuth and altitude angles on a sun path diagram superimposed on a view (in practice, this could be a real panoramic photograph)

Source: Author

and the fraction of this area below the horizon line is termed S, then the shading factor, as a percentage of possible daylight, is:

$$SF = S / T \tag{8}$$

This is the percentage of maximum direct solar radiation on unclouded days that may arrive at the spot.

According to a daylight code (BS (British Standard) 8206-2, 1992), the lines in a sun path diagram can be explained as follows: 'the concentric circles on a stereographic sun-path diagram represent angles of elevation above the horizon; the scale and compass points around the perimetre represent orientation. Each of the long curved arcs gives the sun-path, the solar altitude and azimuth, for a particular day; the shorter, converging, lines give the time of day.'

2.6.1 Using a heliodon

A heliodon device simulates the sun and shadow patterns on a building using a scale model placed on it. Heliodons are best for studying sun and shadow patterns but not daylighting effects, as they cannot replicate reflectance patterns from nearby surfaces, including the

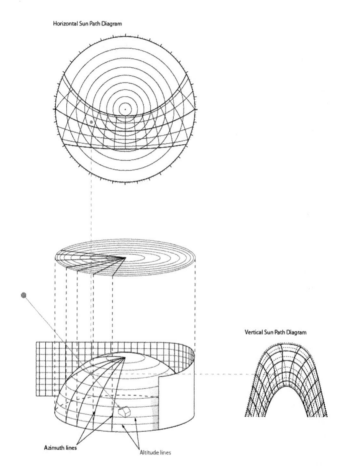

Horizontal Sun Path Diagram

Vertical Sun Path Diagram

Azimuth lines

Altitude lines

Figure 2.7 How horizontal and vertical sun path diagrams relate to the global environment around the building

Source: Author

ground, and conditions in cloudy weather. There are ring heliodons and platform heliodons:

- with a ring heliodon the sun visually tracks across the skyvault;
- the platform heliodon is less intuitive but more accurate.

More information is available from CERES at Ball State University, Muncie, Indiana, USA.[1]

2.6.2 Sunpegs

Sunpegs are simpler than heliodons to build but require several people to operate. A sunpeg device is attached to a model of the building, taken outdoors into direct sunlight and the model is tilted at various angles by hand to achieve desired sun angles. It can also be placed on the floor of a darkened room and a focused-beam light source moved around the model at a sufficient distance to achieve desired sun angles on the model. This is less accurate except at the centre of the beam. Sunpeg devices are constructed from everyday materials using photocopies of diagrams for the proper latitude.[2]

Notes

1 See: http://cms.bsu.edu/academics/centersandinstitutes/ceres/heliodon/staticheliodon
2 See Grondzik, W. T., Kwok, A. G., Ben Stein, B., Reynolds, J. S. *Mechanical and Electrical Equipment for Buildings*, John Wiley and Sons, 1986; Brown, G. Z., *Sun, Wind, and Light: Architectural Design Strategies*, John Wiley and Sons, 2014.

3 Factors affecting building materials choice

An early choice in the design of low-carbon buildings is that of the choice of construction materials. Thermal conductivity, thermal mass and life-cycle carbon impacts are three considerations beyond other architectural and load-bearing factors.

3.1 Thermal conductivity and resistance

Passive buildings require superinsulation in most climates to resist outside temperature extremes. Superinsulation, together with airtightness, also helps to control and moderate internal temperatures and air conditions to keep occupants comfortable. They reduce the need to provide energy for heating and air-conditioning.

The outer skin of the building is externally insulated and thick. Its purpose is to smooth out variations in internal temperature and humidity. To calculate the amount of insulation required and the expected performance of the building it is necessary to know the thermal mass, conductivity and resistance of the construction materials.

Thermal conductivity: known as lambda (λ) or psi or k-value. This tells us how well a material conducts heat. It expresses how much energy it takes to cause one degree of temperature drop through one unit of thickness of the material. It is expressed as:

$$k = Q / T * 1 / A * x / T \tag{1}$$

or the quantity of heat (Q) transmitted over time (t) through a thickness (x), in a direction perpendicular to a surface of area (A), due to a temperature difference (T). The units used are either SI: W/mK or imperial: Btu/(hr × ft × °F). The lower the value the more insulating the material is. To convert one to the other, use the formula 1.730735 Btu/hr × ft × °F = 1 W/mK.

R-value is the thermal resistance of a unit thickness (how much heat passes across a unit thickness of a specific material). It is the ratio of the temperature difference across an insulator and the heat flow per unit area through it. It is equal to the depth/thickness of a material divided by its thermal conductivity – that is,

$$\textbf{R - value} = \frac{l}{k}$$ [2]

where l is the thickness in feet or metres. Units are SI: m²K/W or imperial: ft²F/Btu. The higher the value the more insulating the material is. For materials in series in a wall, roof or floor, R-values are added together to give a thermal resistance for the whole. R-values are frequently cited without units – for example, R-3.5. One R-value (imperial) is equivalent to 0.1761 R-value (SI), or one R-value (SI) is equivalent to 5.67446 R-value (imperial). Usually, the appropriate units can be inferred from the context and their magnitudes.

Example:

1. The thermal conductivity of wood fibre batts insulation is 0.039 W/mK.
2. The thermal conductivity of brick is 0.69 W/mK.
3. The thermal resistance of 210mm of wood fibre batts insulation is 0.021/0.039 = 0.538 m²K/W.
4. The thermal resistance of 220mm of brick is 0.69 W/mK is 0.022/0.69 = 0.032 m²K/W.
5. The thermal resistance of the two combined is 0.032 + 0.538 = 0.571 m²K/W.

U-value is the thermal conductivity of a unit thickness, a measure of how much heat is lost through a given thickness of any specific material, including conduction, convection and radiation expressed as the amount of heat in joules (J) that travels in 1 second (i.e. watts) through 1m² of wall from one side to the other to deliver a 1°C difference in temperature. The unit is SI: W/m²K (watts per hour per square metre per degree Kelvin) or imperial: BTU/ft²F (BTUs per hour per square foot per degree Fahrenheit). The lower the value the more insulating the material is. The U-value of a material (or several materials together) is the reciprocal of the R-value (i.e. 1/R-value), plus convection and radiation heat losses, plus repeating thermal bridges (e.g. wall ties). This is a complex calculation, so U-value calculation software is recommended (see Chapter 10).

> *To convert simple R-values to U-values*: R-value = 1/U-value
> *To convert simple U-values to R-values*: U-value = 1/R-value

Example: Using the values from the previous example, the thermal transmittance of 220mm of brick plus 210mm of wood fibre batts insulation is 1/0.571 = 1.751 W/m²K.

In addition, calculations also use the following data:

- R_{si}: the interior surface heat transfer coefficient;
- R_{se}: the exterior surface heat transfer coefficient.

These differ depending on whether the heat is travelling horizontally, vertically or downwards. The values are found embedded in look-up tables in software such as the Passive House Planning Package (PHPP).

The U-value of a wall is calculated from the formula:

$$\text{U-value} = \frac{1}{(R_{si} + d/\lambda + R_{se})} \qquad [3]$$

Example: A solid concrete wall has the following qualities:

- R_{si}(the interior surface heat transfer coefficient) horizontal = 0.13
- d (depth) = 0.150m
- λ (thermal conductivity) = 1.8
- R_{se} (the exterior surface heat transfer coefficient) horizontal = 0.04

The U-value is therefore:

$$\frac{1}{(0.13 + 0.150/1.8 + 0.04)} = 3.947368 \text{ W/m}^2\text{K}$$

As layers are added to the wall, of insulation, render, etc., then d_2/λ_2 + d_3/λ_3 + ... d_n/λ_n would be added to the series below the line for each material and its depth.

Table 3.1 k-value and depth of insulation required to reach a U-value of 0.15W/m²K[1]

Material	k-value	Depth (mm)
Structural insulated panels	c.0.040	215
Polyurethane with pentane up to 32kg/m³	0.027–0.03	140–50
Polyurethane soy-based	0.026–0.038	130–95
Foil-faced polyurethane with pentane up to 32kg/m³	0.020	110
Polyurethane with CO_2	0.035	170
In-situ applied polyurethane foam (sprayed or injected)	0.023–0.028	120–40
Polyisocyanurate up to 32kg/m³	0.025–0.028	125–40
Foil-faced polyisocyanurate up to 32kg/m³	0.022–0.023	120–5
In-situ applied polyisocyanurate (sprayed)	0.023–0.028	120–40
Phenolic foam (PF)	0.020–0.025	110–30
Foil-faced phenolic foam	0.020–0.023	110–25
Expanded polystyrene (EPS) up to 30kg/m³	0.030–0.045	145–250
Expanded polystyrene with graphite (grey)	0.030–0.032	145–60
Extruded polystyrene (XPS) with CO_2	0.025–0.037	125–90
Extruded polystyrene with HFC 35kg/m³	0.029–0.031	145–60
Glass wool [up to 48kg/m³]	0.030–0.044	150–230
Glass wool [equal/greater than 48kg/m³]	0.036	180
Mineral wool [less than 160kg/m³]	0.034–0.038	170–95
Mineral wool [160kg/m³]	0.037–0.040	190–210
Sheep's wool [25kg/m³]	0.034–0.054	170–330
Cellulose fibre [dry blown 24kg/m³]	0.035–0.046	175–270
Polyester fibre	0.035–0.044	175–260
Wood fibre batts or rolls	0.039–0.061	195–350
Hemp-lime (monolithic)	0.067	380
Cotton	0.039–0.040	195–210
Cork slab [120kg/m³]	0.041–0.055	215–330
Vermiculite	0.039–0.060	195–350
Perlite (expanded) board	0.051	310
Cellular glass (CG)	0.038–0.050	190–320
Flexible thermal linings	0.040–0.063	210–360
Strawboard [420kg/m³]	0.081	450
Straw bale (monolithic)	0.047–0.063	310–60
Hempcrete	0.12–0.13	640

Table 3.2 k-values of other materials[2]

Substance	$\lambda/(W\ m{-}1\ K{-}1)$
Bitumen	0.17
Brick (dry)	0.8–1.2
Concrete:	
cellular	0.1–0.2
lightweight aggregate	0.2–0.6
dense	0.6–1.8
Cork, baked slab	0.038–0.046
granular	0.04
Felt	0.04
Hardboard	0.125
Glass, borosilicate crown	1.1
double extra dense flint	0.55
Glass, light flint	0.85
Pyrex	1.1
Plasterboard (gypsum)	~0.16
Plywood	0.125
Rubber, cellular	~0.045
natural	~0.15
silicone	0.25–0.4
Sand, silver	0.3–0.4
Soil, clay	~1.1
Timber	0.14–0.17

3.2 Thermal mass

Thermal mass is the product of the specific heat capacity of the material and its total mass and conductivity, and can be worked out precisely. Higher thermal mass is indicated by higher heat transfer coefficient, as measured in W/m^2K (watts per square metre Kelvin) as follows:

$$h = \Delta Q\ /\ A \times \Delta T \qquad\qquad [4]$$

where:

h   = heat transfer coefficient, W/m^2K

ΔQ = heat input or heat lost, in watts

A   = heat transfer surface, in m^2

ΔT = difference in temperature between the solid surface and the adjacent air space.

The thermal mass of the building fabric is the sum of the thermal mass of all the building's components. It can be calculated using the values in Table 3.3.

Table 3.3 Thermal mass of building materials

	Density [kg/m³]	Specific heat capacity [J/kg/K]	Thermal conductivity [W/m/K]
Masonry materials:			
sandstone	2,300	1,000	1.8
brick (exposed)	1,750	1,000	0.77
brick (protected)	1,750	1,000	0.56
no-fines concrete	2,000	1,000	1.33
concrete block (dense) (exposed)	2,300	1,000	1.87
concrete block (dense) (protected)	2,300	1,000	1.75
precast concrete (dense) (exposed)	2,100	1,000	1.56
precast concrete (dense) (protected)	2,100	1,000	1.46
cast concrete	2,000	1,000	1.33
cast concrete	1,800	1,000	1.13
lightweight aggregate concrete block	600	1,000	0.2
autoclaved aerated concrete block	700	1,000	0.2
autoclaved aerated concrete block	500	1,000	0.15
screed	1,200	1,000	0.46
ballast (chips or paving slab)	1,800	1,000	1.1
Surface materials/finishes:			
external render (lime, sand)	1,600	1,000	0.8
external render (cement, sand)	1,800	1,000	1
plaster (dense)	1,300	1,000	0.57
plaster (lightweight)	600	1,000	0.18
plasterboard (standard)	700	1,000	0.21
plasterboard (fire-resisting)	900	1,000	0.25
Insulation materials:			
mineral wool (quilt)	12	1,030	0.042
mineral wool (batts)	25	1,030	0.038
expanded polystyrene (EPS)	15	1,450	0.04
extruded polystyrene	40	1,400	0.035
polyurethane foam	30	1,400	0.025
urea formaldehyde (UF) foam	10	1,400	0.04
blown fibre	12	1,030	0.04
Miscellaneous materials:			
plywood sheathing	500	1,600	0.13
timber studding	500	1,600	0.13

(Continued)

Table 3.3 Thermal mass of building materials (*Continued*)

	Density [kg/m³]	Specific heat capacity [J/kg/K]	Thermal conductivity [W/m/K]
timber battens	500	1,600	0.13
timber decking	500	1,600	0.13
timber flooring	500	1,600	0.13
timber flooring (hardwood)	700	1,600	0.18
chipboard	600	1,700	0.14
vinyl floor covering	1,390	900	0.17
waterproof roof covering	110	1,000	0.23
wood blocks	600	1,700	0.14
floor joists	500	1,600	0.13
cement-bonded particle board	1,200	1,500	0.23
carpet/underlay	200	1,300	0.6
steel	7,800	450	50
stainless steel	7,900	460	17
soil	1,500	1,800	1.5

Source: Arup

A free tool for calculating the thermal properties of construction elements is available from http://bit.ly/2hUU6P2. The methodology is based on British Standard BS EN ISO 13786.

3.2.1 Thermal capacity

All building materials have three essential thermal properties:

1. *thermal conductivity*: how well it conducts heat;
2. *thermal mass* (thermal capacity): how much heat energy is required to raise its temperature;
3. *thermal inertia*: how quickly it responds to temperature changes.

It is the combination of these properties that dictates how a material performs in real buildings with changing conditions.

The free tool also calculates thermal capacity, the product of mass × specific heat capacity. Also known as thermal admittance (Y), this is a measure a material's ability to absorb heat from, and release it to, a space over time. It is an indicator of the thermal storage capacity (thermal mass) of a material, absorbing heat from and releasing it to a space through cyclical temperature variations, thus evening out temperature variations and so reducing the demand on building services systems. Thermal admittance is expressed in $W/(m^2K)$, where the higher the admittance value, the higher the thermal storage capacity.

It is calculated as the heat transfer (in watts) divided by area multiplied by the temperature difference between the surface of the material and the air:

$$Y = \frac{W}{(m^2 K)}$$

[5]

The free tool also calculates attenuation gains (thermal diffusivity) and the thermal phase shift (also known as decrement). These are terms for the rate of the passage of heat from one side to the other (see section 3.4 below for more information).

The equation relating thermal energy to thermal mass is:

$$Q = C_{th} \Delta T$$

[6]

where Q is the thermal energy transferred, C_{th} is the thermal mass of the body and ΔT is the change in temperature. Ideal materials for thermal mass if heat needs to be retained in a minimal volume are those materials with both high specific heat capacity and high density.

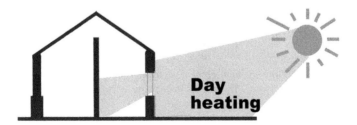

Figure 3.1 Thermal mass in a building may be used to absorb heat in the daytime

Source: Author

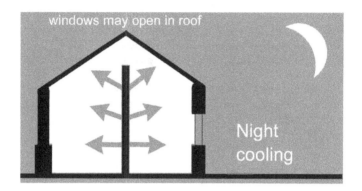

Figure 3.2 Thermal mass in a building used to release heat in the night

Source: Author

3.3   Thermal mass and building design

The choice of a thermally light or heavy building is an important design choice. It affects how much energy is required to maintain a satisfactory indoor climate.

3.3.1   *Thermally light construction*

In a 'thermally light' building the building will cool down quicker, particularly if the air is allowed to leave. In a cool climate, although more energy is required to keep the temperature constant it will warm up more quickly from cold. This might be appropriate if the building is not occupied during the day. Examples include:

- straw bale;
- timber frame with cellulose insulation;
- steel frame with PIR or SIPs insulation.

Advantages:

- ease of construction;
- lower ecological footprint;

- amenable to off-site fabrication;
- small buildings do not necessarily need foundations: for structures up to three storeys the frame sits on pads, stones or rammed earth-filled tyres.

Considerations:

- low thermal buffering (storage);
- thermal bridges must be eliminated in the case of timber and steel frames.

3.3.2 *Thermally massive construction*

In cool weather/climates, a 'heavy' building takes longer to warm up but stays warm for longer because the more dense a material, the more heat it will hold. Thermally massive materials absorb heat and take several days to release it slowly as they cool down. In hot weather/climates thermally massive buildings help to prevent overheating by absorbing internal heat and a certain amount of humidity. In both cases external insulation is necessary to protect the interior temperature. Their degree of success is increased in cases where there are large temperature variations between day and night. Examples of thermally massive construction materials include:

- concrete, hempcrete and substitutes;
- brick and stone;
- rammed earth or cob.

Advantages:

- storage of thermal energy;
- long-lasting.

Considerations:

- potentially higher embodied energy;
- need external insulation to prevent thermal bridging.

3.4   Thermal diffusivity

Thermal diffusivity is a related property to thermal mass: it is a measure of the speed at which the material conducts heat, say, from the outside (when the sun shines on it) to the inside. For protection against internal overheating, insulation materials are chosen that have a low thermal diffusivity and a high thermal insulation level. Thermal diffusivity is defined by:

$$a = \frac{k}{\rho \cdot c}$$

[7]

where:
a = thermal diffusivity (basic SI unit m^2/s; may be given as cm^2/h)
k = thermal conductivity
c = specific heat capacity.
ρ = density (SI unit (kg/m^3)

Materials with good thermal diffusivity display a high amplitude dampening effect and good phase shifting. These are explained next.

Table 3.4 Thermal properties of common building materials

Material	Density ρ (kg/m^3)	Thermal conductivity k (W/m^2K)	Specific heat capacity (J/kg^2K)	Thermal diffusivity a (cm^2/h)
Spruce, pine, fir	600	0.130	2,500	3
Wood fibre sarking and sheathing board	270	0.051	2,100	3
Wood fibre external insulation	250	0.051	2,100	3
Wood fibre rigid insulation	160	0.041	2,100	4
Wood fibre flexible insulation	45	0.039	2,100	15
Hemp flexible insulation	40	0.040	1,700	21
Brickwork	1800	0.800	1,000	16
Reinforced concrete	2200	1.400	1,050	22
Polystyrene	40	0.040	1,380	26
Polyurethane foam	30	0.030	1,380	26
Glaswool	30	0.035	800	52
Steel	7800	58	600	446
Aluminium	2700	200	921	2895

3.4.1 Amplitude dampening

Amplitude dampening is a measure of resistance to temperature penetration. It is also known as temperature amplitude ratio (TAR). It is described by the relationship between the maximum fluctuation of external compared to internal temperatures. If the external temperature fluctuates between -5°C and +30°C (a difference of 35°) and the internal temperature fluctuation is to be between 15°C and 20°C (a difference of 5°), then the amplitude dampening factor is 35/5 = 7.

3.4.2 Phase shifting

Phase shifting denotes the time taken for an extreme external temperature to reach the interior. The aim of passive cooling in hot, non-humid areas is to have a phase shift of 12 hours, so that the midday heat only reaches the interior in the middle of the night. With good amplitude dampening, its effect will also be very much moderated. Some of the stored energy in the fabric is thus automatically transferred back outside, ensuring that temperature fluctuations and extremes on the inside are much less than outdoors.

Amplitude dampening and phase shifting should be particularly observed in roof areas. These may get very hot and so require large thicknesses of insulation with low thermal diffusivity. Aim for an amplitude dampening factor of 10 and a minimum phase shift value of 10 hours.

Example 1: a roof insulated with mineral wool with a thermal conductivity of 0.035, density 20 kg/m^3 and U-value 0.18 W/m^{2}K will achieve an amplitude dampening factor of 6 and a phase shift of 6.8 hours. If the midday temperature is 35°C, at 10 pm it will be 29°C beneath the roof.

Example 2: a roof insulated with wood fibre insulation with the same thermal conductivity, but a density of 50 kg/m^3 (five times greater thermal storage mass) gives an amplitude dampening factor of 12 and a phase shift of 11 hours. Therefore at 1am it will be 21°C beneath the roof. Night ventilation will reduce the temperature further.

Materials with a lower thermal diffusivity will conversely have the effect that interiors, once heated, will take longer to cool down.

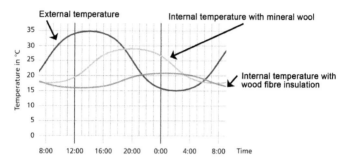

Figure 3.3 Effect of insulation on thermal phase shift

Source: Steico/author

3.5   Carbon intensity

It is self-defeating for mitigating climate change to construct a passive solar building of materials whose manufacture causes significant greenhouse gas emissions. Conversely, it is ideal to use materials which lock up atmospheric carbon dioxide, when well-performing and suitable alternatives exist. The possibility then exists of constructing a carbon-negative building. Natural cellulose-based materials offer this possibility. Steel is a possible exception, despite its high embodied carbon from fossil fuels, as it is long-lasting and can be recycled, but many large and high-rise structures are now being made using frames made of timber products.

3.6   Cellulose-based materials

'Natural', 'green', 'bio' or 'renewable' building materials can be classed together as 'cellulose-based'.[3] Among their benefits are that they:

- lock up atmospheric carbon in the building;[4]
- have varying degrees of insulation ability;
- are easy to work with;
- make structures that are breathable.

They are also biodegradable or easily recycled at the end of the building's life and may support local agroforestry.

3.6.1 Wood

Wood has a greater tensile strength relative to steel – two times on a strength-to-weight basis – and has a greater compressive resistance strength than concrete. Sustainably sourced timber must be specified. Products for structural use include glued laminated timber ('glulam') and cross-laminated timber (CLT).

3.6.2 Glued laminated timber ('glulam')

To make this, layers of dimensioned timber ('laminating stock' or 'lamstock' in the USA) are bonded with durable, moisture-resistant structural adhesives to form single large, strong, structural members from smaller pieces. These can produce columns, beams and curved, arched shapes. Glulam has a much less embodied energy per unit mass than reinforced concrete and steel, although more than solid timber. The tallest building in the world made with timber frame is 14 storeys high and is in Bergen, Norway. It uses metre-thick columns of glulam and CLT, plus two concrete decks above the fifth and tenth floors.

The high strength and stiffness of laminated timbers enable glulam beams and arches to span large distances without intermediate columns, allowing more design flexibility than with traditional timber construction. It is typically specified structurally in strength class C24. Glulam has the following characteristics (dependent upon the strength class, of which there are eight):

- thermal conductivity: 0.13 W/mK (through glued joints); 0.13 W/mK (vertical to glued joints);
- water vapour diffusion resistance factor: 40;
- density: 380–480 kg/m^3;
- compressive strength:[5] 2.7–3.3 N/mm^2 (perpendicular to grain of boards); 24–50 N/mm^2 (parallel to grain of boards);
- bending strength:[6] 19–50 N/mm^2 (parallel to grain of boards);
- elastic modulus:[7] 390–460 N/mm^2 (perpendicular to grain of boards); 12,600–18.500 N/mm^2 (parallel to grain of boards).[8]

3.6.3 Cross-laminated timber (CLT)

CLT is an engineered timber product with good structural properties and low environmental impact able to provide dry, fast on-site construction, with good potential for airtightness and a robust wall and floor structure suitable for most finishes internally and externally. It requires only limited new site skills and can be assembled without the use of adhesives using mechanical fixing. Its low weight means that a high degree of off-site manufacture is possible.

Relatively large pre-manufactured panels can be transported to site for fast, quality construction. It is used to form the structural floor and wall element of buildings. Buildings up to nine storeys tall have been successfully constructed in this fashion in the UK. As a natural renewable product performance can vary slightly, but commercial CLT systems generally achieve:

- thermal conductivity: 0.13 W/mK;
- density: 480–500 kg/m^3 (based on spruce timber);
- compressive strength: 2.7 N/mm^2 (perpendicular to grain of boards); 24–30 N/mm^2 (parallel to grain of boards);
- bending strength: 24 N/mm^2 (parallel to grain of boards);
- elastic modulus: 370 N/mm^2 (perpendicular to grain of boards); 12,000 N/mm^2 (parallel to grain of boards).[9]

Other useful timber products include:

- plywood, wood structural panel;
- oriented strand board (OSB);
- laminated veneer lumber (LVL);
- parallel strand lumber (PSL);
- laminated strand lumber (LSL);
- finger-jointed lumber;
- I-joists and wood I-beams – 'I'-shaped structural members designed for use in floor and roof construction;
- roof trusses and floor trusses;
- medium density fibreboard (MDF).

Most of these timber products are bonded with wax and synthetic moisture-resistant resin adhesives. In such cases it is best to avoid the use of phenol formaldehyde (PF) and melamine fortified urea formaldehyde (MUF) as they can emit volatile organic compounds (VOCs) such as formaldehyde and impair internal air conditions. Instead, it is possible to choose products bound with isocyanate (PMDI), a binder which does not emit VOCs. These products can qualify for 'green' home certifications such as LEED® and National Green Building Standard®.

3.6.4   *Eurocodes and timber construction*

Eurocodes are a suite of harmonised European standards developed by the European Committee for Standardisation that are applicable to all construction works across the European Union. The calculations are useful if similar standards are not extant in other territories. They cover the following aspects of construction:

- structural design;
- actions on structures;
- the design of concrete, steel, composite steel and concrete, timber, masonry and aluminium structures;
- geotechnical design;
- design of structures for earthquake resistance.

Eurocode 5 standards specifically relate to timber structures. They include British Standard BS EN 1995, which covers common rules and rules for buildings, structural fire design and bridges; BS EN 1995 uses the limit state concept, not the permissible stress method, and requires the use of software by the designer.[10]

3.6.5   *Other natural carbon-negative materials*

See Table 3.5

3.6.6   *Straw bale*

Straw bale is used as infill in timber frame structures and is rendered with hempcrete or lime. It has also been used structurally: the tallest frameless straw bale building is three storeys high.[12]

Typical properties of straw bale:

- minimum recommended bale dry density: 110–30 kg/m^3;
- thermal conductivity: 0.055–0.065 W/mK (density 110–30 kg/m^3);
- recommended initial moisture content: 10–16 per cent;
- recommended maximum in-service moisture content: normally not to exceed 20–25 per cent.[13]

Table 3.5 Other cellulose-based materials currently used in construction[11]

Material	Application
Flax	Roofing insulation
Hemp fibres	Insulation
	Medium density fibre board
	Oriented strand board
Hemp shiv	Monolithic construction of walls, floors and roofs
	Insulation
	Panel construction
Jute	Carpet
	Plastering mesh
	Scrim
Paper	Recycled and shredded for insulation
	Mixed with cement to form blocks
Reed	Thatching
Reed mats	Plastering base (like laths)
Sisal	Carpet (mixed with reinforced cement in some countries)
Straw	Bales as building blocks
	Wall panels
	Thatching

A 500mm thick structural straw wall with finishes has good insulation: a U-value of around 0.15 W/m^2K.

3.6.7   Natural fibre insulation

Includes insulation products derived from natural products such as wood fibre and cellulose, wool, hemp, cotton and flax. These are often used as replacements for mineral- or petrochemical-based insulation to deliver comparable thermal and acoustic insulation but with a lower or potentially negative carbon footprint and fewer health issues during installation. They can also assist in regulating relative humidity and can provide a vapour-permeable system.

3.6.8   Eco-alternatives to concrete

Concrete is made from varying proportions of coarse aggregate bonded with cement that hardens over time. Most concretes used are lime-based concretes made from calcium silicate such as Portland cement. The main ingredient is limestone or calcium carbonate

Table 3.6 Natural fibre insulations

Material	Typical thermal conductivity (W/m/K)	Commonly available formats
Wood fibre	0.038–0.050	Boards, semi-rigid boards and batts
Paper (cellulose)	0.035–0.040	Loose batts, semi-rigid batts, loose
Hemp	0.038–0.040	Semi-rigid slabs, batts
Wool	0.038–0.040	Semi-rigid boards, rolls
Flax	0.038–0.040	Semi-rigid boards, rolls
Cork	0.038–0.070	Boards, granulated

Source: BRE

($CaCO_3$). Portland cement is made by heating the raw materials, including the limestone, first, to above 600 °C (1,112 °F) and then to around 1,450 °C (2,640 °F) to sinter the materials. This emits carbon dioxide and produces calcium silicate (($CaO)_3 \cdot SiO_2$). The fossil fuel use for heating also emits carbon dioxide. When it is turned into liquid cement with the addition of water and exposed to the air it absorbs carbon dioxide again, to reform into calcium carbonate ($CaCO_3$), and hardens. Recent research[14] has shown that an average of 42 per cent of greenhouse gas emissions associated with the creation of cement are actually recouped from the atmosphere once the concrete is in situ.

Various ways[15] are available to reduce the environmental impact of concrete manufacture; others (including carbon capture) are in R&D phase.

Alternatives to Portland cement also exist which store atmospheric carbon. Some use the waste material fly ash, others use magnesium silicate. These materials react, like hydraulic lime, with the carbon dioxide in the air, absorbing it as it hardens. These alternatives include:

- *hemcrete*, a mixture of lime and hemp shiv, also stores carbon (but its typical compressive strength is around 1MPa, over 20 times lower than low grade concrete; density: 15 per cent that of traditional concrete; k-value: 0.12–0.13 W/mK). Like other plants, the hemp absorbs carbon from the atmosphere as it grows. 165 kg of carbon can be theoretically absorbed and locked up by 1m³ of hemp-lime wall over many decades;

- *Calera* and *SOLIDIA* calcined products;
- *magnesium and alumina silicate*-based crete;
- *Celitement*, or calcium hydrosilicate;
- Fly ash cement;
- *Ferrock*, composed partly of iron dust reclaimed from steel mills and currently sent to landfill;
- *Aether*, a belite-calcium sulfo-aluminate-ferrite compound;
- Canada's *CarbonCure* Technologies' low-carbon concrete masonry.

All of these alternatives are comparatively expensive as yet. Stone, unfired clay masonry and rammed earth are traditional alternative materials.

Hemp-lime walls must be used together with a timber frame to support the vertical load of the building. Hemp-lime can either be applied by hand or sprayed on in layers, allowing previous layers to dry out. It is vital to get the lime–shiv ratio right, otherwise there will be problems with drying out. If less binder (lime) is added the density decreases and the insulation U-value improves, and vice versa.

3.6.9   *Unfired clay masonry*

Unfired clay masonry may be considered as an alternative to concrete and concrete blocks. A low-impact and traditional building material, it is a relatively robust, fire-resistant material, with good thermal mass and the ability to moderate internal humidity levels. Unless stabilised, unfired clay masonry is not resistant to prolonged water exposure and should normally be protected from rain. Once moulded, unfired bricks are left to dry in controlled conditions, rather than fired, which significantly reduces their overall embodied energy. Typical properties of commercially supplied unfired clay block systems are:

- dry density: 1,700–2,200 kg/m^3;
- compressive strength (depends on moisture content): 1–4 N/mm^2;
- thermal conductivity (depends on density): 0.5–1.0 W/mK.[16]

(As a natural product, performance can vary slightly.)

3.7   Estimating the carbon content of buildings

For reporting purposes, it may be useful to report the carbon storage capacity of a building structure, to add to the carbon saved from using renewable energy to heat and power it. Table 3.7 may prove helpful for wood quantities when known, but for buildings a web-based calculator[17] may be used. For other materials, a rough estimate would be to take a softwood (commonly used for natural insulation) and divide the carbon content by the difference in density. For example, for straw bale (see below), this has 120/492 = 24.4% the density of spruce. Therefore, it may be assumed to sequester 24.4% × 193 = 47.07kg/m^3 of carbon, using the information in Table 3.7 (see p. 48).

Notes

1 Energy Saving Trust.
2 http://www.kayelaby.npl.co.uk/general_physics/2_3/2_3_7.html and http://www.engineeringtoolbox.com/thermal-conductivity-d_429.html
3 See *Cellulose-based building materials*, NHBC Foundation, January 2014.
4 An average new house made with conventional materials contains the equivalent of 50 tonnes of CO_2 as 'embodied carbon' that could be reduced to 38 tonnes with greater use of timber and modern methods of construction, or even to approximately 25 tonnes by using cellulose-based materials, such as hempcrete. See Monahan, J. and Powell J.C., 'An embodied carbon and energy analysis of modern methods of construction in housing: a case study using a lifecycle assessment framework', *Energy and Buildings*, 2011, 43: 179–88.
5 Compressive strength is the maximum compressive stress that, under a gradually applied load, a given solid material can sustain without fracture. Compressive strength is calculated by dividing the maximum load by the original cross-sectional area of a specimen in a compression test.
6 Bending strength is a measure of the tensile strength of beams or slabs, or the amount of stress and force an unreinforced slab, beam or other structure can withstand such that it resists any bending failures.
7 Elastic modulus is the measure of rigidity or stiffness of a material. It is a ratio of stress to strain. Plotting these two variables on a graph for a given material under test produces a stress–strain curve. The modulus of elasticity is the slope of the stress–strain curve in the range of linear proportionality of stress to strain. The greater the modulus, the stiffer the material, or the smaller the elastic strain that results from the application of a given stress.
8 Source: Constructional Timber.
9 Source: BRE.

Table 3.7 Wood density and carbon content of commonly used tree species at a kiln-dry moisture content of 15%

	Standard units			Metric units		
	Density*	Estimated carbon content**	Estimated CO_2 equivalent content** (15% MC)	Density*	Estimated carbon content**	Estimated CO_2 equivalent content** (15% MC)
	Pounds per ft³ for kiln-dried wood			Kilograms per m³ for kiln-dried wood (15% MC)		
Cedar, western red	24.6	10.7	39.2	394	171	627
Douglas fir/larch	34.5	15.0	55.0	553	240	880
Hem/Fir	30.7	13.3	48.8	492	214	785
Spruce/Pine/Fir	27.8	12.1	44.4	445	193	708
Pine, southern yellow	36.3	15.8	57.9	582	253	928
Redwood	24.0	10.4	38.1	385	167	612
Red oak	44.5	19.3	70.9	713	309	1,136

Notes: * To determine the dry (moisture-free) weight, divide density numbers shown above (columns one and four) by 1.15
** The carbon content per unit volume is less in green wood (S-GRN). For approximate carbon content of a given volume of green wood, multiply carbon content values above by 0.95
Source: *Carbon in Wood Products: The Basics*, Dovetail Partners, Minneapolis, 2013, available at http://www.dovetailinc.org/land_use_pdfs/carbon_in_wood_products.pdf

10 See http://shop.bsigroup.com/Browse-By-Subject/Eurocodes/Descriptions-of-Eurocodes/Eurocode-5/

11 Yates, T., 'The use of non-food crops in the UK construction industry: case studies', *Journal of the Science of Food and Agriculture*, 86: 1709—96, Chichester, Wiley Interscience, 2006.

12 Rachel Shiamh's house in Ceredigion, Wales, featured as a case study in Thorpe, D., *The 'One Planet' Life: A Blueprint for Low Impact Development*, Routledge, 2015.

13 Source: BRE.

14 Fengming Xi *et al.*, 'Substantial global carbon uptake by cement carbonation', *Nature Geoscience*, 9, 880–3, 2016. doi:10.1038/ngeo2840 available at: http://go.nature.com/2gUb9Qr

15 See EnergyStar guide for energy and plant managers, *Energy efficiency improvement and cost saving opportunities for cement making*, available at http://bit.ly/2hqz9KU and GreenConcrete LCA, a web-based tool based on MS-Excel to analyse environmental impacts of the production of concrete and its constituents, available at http://bit.ly/2hSVvG8

16 Source: BRE.

17 See http://www.woodworks.org/design-and-tools/design-tools/online-calculators/

4   Passive design principles

Depending on the location, climate and weather, heat gains are either welcome or not. Whichever is the case, they need to be calculated and factored into the building's design.

4.1   Heat gains in buildings

Heat gains arise from the following sources: occupants' body heat; specific heating systems; cooking; clothes cleaning and drying; output from electrical/electronic devices and equipment; use of hot water; outside temperature conducted through the roof, walls and floor; direct solar insolation through the glazing; air movement through the building.

A method for establishing the design conditions begins by determining the indoor and outdoor temperature and wind conditions. ASHRAE[1] guidelines specify using the design dry bulb temperature and outdoor summer design temperature for almost all basic heat gain calculations, but, ideally, the building should be designed to function optimally by adapting to all seasonal requirements. An alternative treatment is given in section 7.5 on the Passive House Planning Package (PHPP).

4.1.1   Internal heat gains

Heat gains arising from loads for each hour of the year are estimated on the basis of per cent of peak design load as they vary from hour to hour and during the year. They can be estimated on the basis of the type of building, and this information may be available as benchmark figures based on a particular region, country, economy and society. Heat arising may be latent (from people and equipment in moisture or water vapour) or sensible (from people, lights and equipment).

Latent heat is an instantaneous cooling load. Sensible heat is a time-delayed cooling load.

As with solar heat entering the space, part of sensible heat generated by internal sources is first absorbed by the surroundings and then released gradually. To allow for the time delay due to thermal storage, cooling load factors (CLFs) have been developed to estimate the heat gains from internal heat-emitting sources. These are based on the time (hour) when the internal source starts to generate heat load and the number of hours it remains in operation. This information is expressed as hourly internal load profiles (percentage of design).

People

The heat gain from people can become significant. The following is a summary of the methodology. To calculate the heat gain from people, use the expected (design) occupancy density for the building. The heat arising from people is calculated by:

$$Q\text{-}ps = N\text{-}p \cdot Fu \cdot qs \cdot CLF\text{-}h \text{ (sensible heat gain)} \qquad [1]$$

$$Q\text{-}pl = N\text{-}p \cdot Fu \cdot ql \text{ (latent heat gain)} \qquad [2]$$

where:
Q-ps = sensible heat gain (SHG) from people
Q-pl = latent heat gain (LHG) from people
 N-p = number of people (maximum or design from occupancy criteria for building)
   Fu = diversity factor or percentage of maximum design for each hour of the day
      = 0 when there are no people in the room
      = 1 when the maximum design number of people are in the room $0 \leq Fd \leq 1$
   qs = SHG per person for the degree or type of activity in the space,[2] e.g. 0.072kW (245 Btu/hr) per person when working in an office and 0.170kW (580 Btu/hr) per person performing heavy manual work in a factory
   ql = LHG per person for the degree or type of activity in the space, e.g. 0.045kW (155 Btu/hr) per person when working in an office and 0.255kW (870 Btu/hr) per person performing heavy manual work in a factory

CLF-h = CLF for given hour. This depends on zone type, hour entering space and number of hours after entering into space.[3]

Lights

The heat gain from lights is calculated from:

$$Q\text{-}l = W \cdot Fu \cdot Fs \cdot CLF\text{-}h \text{ (sensible heat gain)} \qquad [3]$$

where:
   Q-l = sensible heat gain from lights
   W = lighting power output in watts (to convert to Btu/hr multiply by 3.412)
  Fu = usage factor or percentage of maximum design for each hour of the day
     = 0 when all lights are off
     = 1 when the maximum design number lights are on
       0 <= Fu <= 1, e.g. Fu = 0.5 when 50% of lights are on
  Fs = service allowance factor or multiplier (accounts for ballast losses in fluorescent lights and heat returned to return air ceiling plenum in the case of air-light fixtures)
CLF-h = CLF for given hour. This depends on zone type, total hours that lights are on and number of hours after lights are turned on. The sensible heat is absorbed by the surroundings and then released into the air. The CLF accounts for this time delay.

Table 4.1 Examples of sensible heat gain and latent heat gain from people

Level of activity	Typical application	Heat gain/person (kW/Btuh)	
		SHG (qs)	LHG (ql)
Seated at rest	Theatre	0.072/245	0.031/105
Seated, light work	Office	0.072/245	0.045/155
Moderate office work	Office	0.073/250	0.059/200
Standing, walking slowly	Retail sales	0.073/250	0.073/250
Light bench work	Factory	0.081275	0.139/475
Dancing	Nightclub	0.89/305	0.160/545
Heavy work	Factory	0.170/580	0.255/870

Equipment

Equipment consists of three types:

1. electric resistance sensible load (e.g. electric cooker);
2. electric inductive sensible load (e.g. motor);
3. sensible and latent loads (e.g. electric or gas kettle).

To calculate equipment sensible heat gain:

$$Q\text{-es} = W \cdot Fu \cdot Fp \cdot CLF\text{-h} \qquad [4]$$

where:

W = equipment output in watts

Fu = usage factor or percentage of maximum design for each hour of the day

= 0 when all equipment is off

= 1 when the maximum design number of equipment is on

$0 <= Fu <= 1$, e.g. $Fu = 0.5$ when 50% of equipment is on

Fp = part load operating factor for motor type, e.g. a compressor operating at 50% capacity might still use 80% of electric power

P = rated electrical power of equipment motor

Eff = motor efficiency

$CLF\text{-h}$ = CLF for given hour. This depends on zone type, total hours that lights are on and number of hours after lights are turned on. The sensible heat is absorbed by the surroundings and then released into the air. The CLF accounts for this time delay.

To calculate equipment latent heat gain:

$$Q\text{-el} = Mw \cdot Hfg \cdot Fu \cdot Fp \qquad [5]$$

where:

$Q\text{-el}$ = latent heat gain from equipment

Mw = mass of water converted to steam (evaporated or boiled)

Hfg = heat required to convert that water to steam.

The latent heat from equipment such as kettles and dishwashers is an instantaneous cooling load. CLFs do not apply to latent loads.

To calculate the total cooling load the following information will then need to be taken into account: latitude; hourly occupancy levels by time of day and by internal zone of the building (north, south, east, west and central); where the solar heat gain for each zone peaks at different times. If, say, it is an office or work space, assume that the design amount of people/lighting/equipment use commences in each space at the start and end of the working day.

In reality, the per hour occupancy levels and usage will vary and this must be factored in also. For each of the four sides, roof and floor of the building, the cooling/heating loads are calculated by the hour. The CLFs are applied. From the U-values (R-values) of the building envelope the total cooling load is then calculated. Degree days (see section 4.5) are then applied to determine the effect of solar gain.

A worked example can be found at: http://bit.ly/2sgxVeJ

4.2   Windows

Windows admit both light and heat. They also may lose heat through air leakage, conduction and radiation. In passive buildings windows will have two or three layers of glazing and multiple seals, plus internal insulation to prevent thermal bridging. Passivhaus standard windows are typically triple glazed in higher latitudes. Panes of glass will be treated with special coatings that admit and retain as much infrared heat and light as is required. A huge range of coatings is available. Panes may be specified with extra clear outer layers, letting up to 80 per cent of light and 71 per cent of the sun's heat in, for high latitudes, or tinted for low latitudes. The thermal resistance of air each side of the glass of a window contributes most significantly to its thermal resistance. Still air is a good insulator.

4.2.1   Window energy labels

Modern windows are rated by national bodies and come with a declaratory label. In the UK this is the British Fenestration Rating Council (BFRC). This label displays the following information:

1.   *the rating level*: A, B, C, etc.;
2.   *the energy rating*: e.g. −3kWh/(m²K)/yr (= a loss of three kilowatt hours per square metre per year);

Occupancy schedule (profile)

Weekday hour	1	2	3	4	5	6	7	8	9	10	11	12	13	14	15	16	17	18	19	20	21	22	23	24
Occupancy (%)	0	0	0	0	0	5	10	20	95	95	95	95	50	95	95	95	95	30	10	5	0	0	0	0

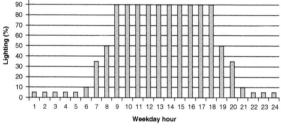

Lighting schedule (profile)

Weekday hour	1	2	3	4	5	6	7	8	9	10	11	12	13	14	15	16	17	18	19	20	21	22	23	24
Lighting (%)	5	5	5	5	5	10	35	50	90	90	90	90	90	90	90	90	90	90	50	35	10	5	5	5

Equipment schedule (profile)

Weekday hour	1	2	3	4	5	6	7	8	9	10	11	12	13	14	15	16	17	18	19	20	21	22	23	24
Receptacles (%)	0	0	0	0	0	10	20	50	90	90	90	90	50	90	90	90	90	70	50	30	20	5	0	0

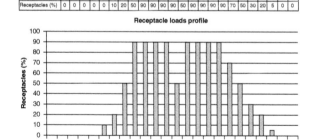

Figure 4.1 Examples of usage factors (Fu), sometimes referred to as operating schedules or profiles. These are used in the hourly calculations above

3. *the U-value*: e.g. $1.4W/(m^2K)$;
4. *the effective heat loss due to air penetration through the edges of openable windows*: e.g. $0.01W/(m^2K)$;
5. *the solar heat gain G-value*: e.g. 0.43 (see section 4.2.2).

The ratings are ranked according to the amount of thermal transmittance (heat transfer) that is permitted by the window, as measured by the amount of energy (in kilowatt hours) lost per year through one square metre of the window, divided by the difference in temperature between the inside and outside $(kWh/(m^2K)/y)$. See Table 4.2.

In the USA, the equivalent label (see Figure 4.2) is produced by the National Fenestration Rating Council (NFRC). Its Component Modeling Approach (CMA) Product Certification Program enables whole product energy performance ratings for commercial (non-residential) projects. It uses the following components:

*   *glazing*: glazing optical spectral and thermal data from the International Glazing Database (IGDB);
*   *frame*: thermal performance data of frame cross-sections;
*   *spacer*: K_{eff} (effective conductivity) of spacer component geometry and materials.

Most US states' building energy codes reference NFRC 100 and 200 for fenestration U-factor and SHGC because they are required by ASHRAE 90.1, Section 5.8.2. California's Title 24 Building Energy Efficiency Standard now requires CMA Label Certificates for site-built fenestration in large projects.

Table 4.2 Meaning of the BFRC window energy label rating

Rating	Energy lost per year (kWh/m^2K)
A	0 (no energy lost) or better
B	0 to −10
C	−10 to −20
D	−20 to −30
E	−30 to −50
F	−50 to −70
G	−70 or worse

National Fenestration
Rating Council®

CERTIFIED

World's Best Window Co.

Millennium 2000+
Vinyl-Clad Wood Frame
Double Glazing • Argon Fill • Low E
Product Type: **Vertical Slider**

ENERGY PERFORMANCE RATINGS

U-Factor (U.S./I-P)	Solar Heat Gain Coefficient
0.30	**0.30**

ADDITIONAL PERFORMANCE RATINGS

Visible Transmittance	Air Leakage (U.S./I-P)
0.51	**0.2**

Manufacturer stipulates that these ratings conform to applicable NFRC procedures for determining whole product performance. NFRC ratings are determined for a fixed set of environmental conditions and a specific product size. NFRC does not recommend any product and does not warrant the suitability of any product for any specific use. Consult manufacturer's literature for other product performance information.
www.nfrc.org

Figure 4.2 A sample window rating label from the National Fenestration Rating Council (NFRC). It gives the solar heat gain coefficient (SHGC) for the glazing, the U-factor (the same as the R-value, or inverse of the U-value), air leakage rate and its visible transmittance (VT)

Source: NFRC

4.2.2   Angle of incident radiation upon glass and shading coefficients

The amount of solar gain and light transmitted through glass is affected by the angle of incidence at which it hits the window. If within 20 degrees of the perpendicular it will mostly pass through it; at over 35 degrees the majority of the energy will reflect off.

Optimum window designs for the available wall can be modelled using a photographic light meter and a heliodon (which adjusts the angle between a flat surface and a beam of light to match the angle between a horizontal plane at a specific latitude and the solar beam). Software is available to perform the same task and calculate cooling and heating degree days (see Chapter 10), and therefore energy performance, using regional climatic conditions available from local weather services.

Shading coefficients are used to measure the solar energy transmittance through windows. The coefficient figure is multiplied by the amount of incoming solar radiation to calculate the proportion that is transmitted through. In Europe this coefficient is called the 'G-value', while in North America it is 'solar heat gain coefficient' (SHGC). G-values and SHGC values ranges from 0 to 1, a lower value representing less solar gain. Shading coefficient values are

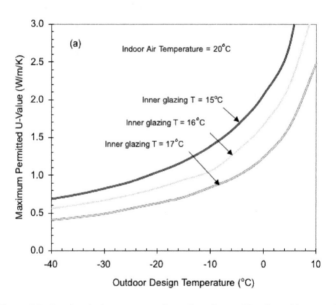

Figure 4.3 Amplitude dampening and window choice. Use this table to read off the window U-value required to yield a given minimum interior temperature compared to the outside temperature

Source: Harvey and Siddall, 2008[4]

calculated using the sum of the direct solar transmittance (T-value) and the secondary transmittance as shown in Figure 4.2.

Primary transmittance is the fraction of solar radiation that directly enters a building through a window compared to the total amount of solar radiation (insolation) that the window receives.

The secondary transmittance is the fraction of inwardly transmitted solar energy absorbed in the window (or shading device), again compared to the total solar insolation.

4.3   Heat flows in buildings

Heat flow into or out of a building occurs through walls, windows, doors, ceilings, floors and openings/gaps by transmission, ventilation and infiltration. It may be a negative figure if the exterior is warmer than the interior. A simple way of calculating the overall loss for a temperate zone climate is[5]:

$$H = H_t + H_v + H_i \tag{6}$$

where:
H = overall heat flow (W)
H_t = heat flow due to transmission through building envelope (W)
H_v = heat flow through ventilation (W)
H_i = heat flow due to infiltration (W).

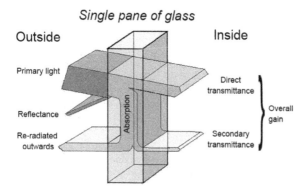

Figure 4.4 Aspects of the SHGC

Source: Author

H_t can be calculated as:

$$H_t = 1.15 * A \, U \, (t_i - t_o) \qquad [7]$$

where:
H_t = transmission heat flow (W)
A = area of exposed surface (m²)
U = overall heat transmission coefficient (W/m²K)
t_i = inside air temperature (°C)
t_o = outside air temperature (°C).

The multiplier 1.15 accounts for additional heat flow through roofs due to absorbed radiation.

If the floors or wall are against earth, the formula becomes:

$$H = A \, U \, (t_i - t_e) \qquad [8]$$

where t_e= earth temperature (°C).

This is a general approach. The use of dedicated software or more detailed evaluation is recommended (see Chapter 10).

4.3.1   Heat flow by ventilation

The heat flow due to ventilation (without heat recovery) can be expressed as:

$$H_v = c_p \, \rho \, q_v \, (t_i - t_o) \qquad [9]$$

where:
H_v = ventilation heat flow (W)
c_p = specific heat capacity of air (J/kg K)
ρ = density of air (kg/m³)
q_v = air volume flow (m³/s)
t_i = inside air temperature (°C)
t_o = outside air temperature (°C).

In temperate/cool weather the heat loss due to ventilation with heat recovery can be expressed as:

$$H_v = (1 - \beta/100) \, c_p \, \rho \, q_v \, (t_i - t_o) \qquad [10]$$

where β = heat recovery efficiency (%).

4.3.2 Heat flow by infiltration

This value is hard to predict and depends on several variables: wind speed, difference between outside and inside temperatures, the quality of the building construction etc. It may be roughly calculated as

$$H_i = c_p \, \rho \, n \, V \, (t_i - t_o) \qquad [11]$$

where:
H_i = heat flow due to infiltration (W)
c_p = specific heat capacity of air (J/kg/K)
ρ = density of air (kg/m^3)
n = number of air shifts, how many times the air is replaced in the room per second (1/s) (0.5 1/hr = 1.4 10^{-4} 1/s as a rule of thumb) (as measured with a blower door test)
V = volume of room (m^3)
t_i = inside air temperature (°C)
t_o = outside air temperature (°C).

An online heat flow calculator can be found in Chapter 10.

4.3.3 Heat flow across building elements

According to Kishore[6] the overall resistance of a single layer wall to heat flow R between the air on either side of an internal surface h_i and an external one h_o is given by:

$$R = 1/h_i + d/k + 1/h_o \qquad [12]$$

The rate of heat flow (q) per unit area from internal air (at temperature T_i) to external air (at temperature T_o) under steady-state conditions can be calculated from:

$$q = \frac{T_i - T_o}{R} \qquad [13]$$

But there are other properties of materials that affect the rate of heat transfer:

- surface characteristics with respect to radiation: absorptivity, reflectivity and emissivity;
- surface coefficient;
- heat capacity;
- transparency to radiation of different wavelengths.

If absorptivity is denoted by a and reflectivity by r, then:

$$r = 1 - a \qquad [14]$$

for any specific wavelength, absorptivity and emissivity are numerically equal.

'Sol-air' temperature is the temperature of air which will result in the same heat flux into a building taking into account the effects of incident radiation. It is defined as:

$$T_{sa} = \frac{T_o + (I \cdot a)}{h_o} \qquad [15]$$

where:
T_{sa} = sol-air temperature
T_o = outside air temperature
I = radiation falling on the plane of the surface (W/m²)
a = absorptivity of the surface
h_o = outside heat transfer coefficient (W/m²K).

Understanding of quantitative heat flows is extremely important in designing passive buildings. The reader is referred to proprietary software designed for this purpose, especially since conditions, in practice, fluctuate constantly (See Chapter 10).

4.3.4 Thermal bridging

Thermal bridging occurs when a conductive material is present through the building envelope from the interior to the exterior. This may be a single-paned window or a concrete lintel in a windowsill, any fixings, window or door frames, joists, services and wall ties in cavities. Sometimes the floor slab is extended through the building envelope.

Whatever it is, it conducts heat undesirably from the inside to the outside or vice versa. As the standard of airtightness and insulation increases in a building, thermal bridging becomes increasingly significant as a factor in heat transfer. To 'break' a thermal bridge, insulation is added on the inside or outside, or even in between, as used in Passivhaus door and window frames.

4.4   Human thermal comfort

In any design for a building the comfort of occupants must be taken into consideration. Thermal comfort is calculated as a heat transfer energy balance. A method of describing thermal comfort was developed by Ole Fanger and is termed predicted mean vote (PMV) and predicted percentage of dissatisfied (PPD). The term 'vote' is used because the analysis was made from the opinions of individuals placed in the particular situations.

4.4.1   Predicted mean vote

The predicted mean vote (PMV) is a thermal scale that runs from Cold (−3) to Hot (+3). It is an ISO standard[7].

The recommended acceptable PMV range for thermal comfort from ASHRAE 55[8] is between −0.5 and +0.5 for an interior space. An online widget may be used to explore the factors that play into human comfort (available at http://bit.ly/1vJDmrL). See Chapter 10 for an online tool to calculate PMV.

Table 4.3 The predicted mean vote sensation scale

Value	Sensation
−3	Cold
−2	Cool
−1	Slightly cool
0	Neutral
1	Slightly warm
2	Warm
3	Hot

4.4.2 Predicted percentage of dissatisfied

Predicted percentage of dissatisfied (PPD) predicts the percentage of occupants that will be dissatisfied with the thermal conditions. The recommended acceptable PPD range for thermal comfort from ASHRAE 55 is <10 per cent persons dissatisfied.

PMV is widely used. The ISO Standard 7730 (ISO 1984), 'Moderate Thermal Environments: Determination of the PMV and PPD Indices and Specification of the Conditions for Thermal Comfort', uses limits on PMV as an explicit definition of the comfort zone. The PMV equation applies to individuals exposed for a long period to constant conditions at a constant metabolic rate:

$$
\begin{aligned}
PMV = [0.303e^{-0.036M} + &0.028]\{(M-W) - 3.96E^{-8} \\
&f_{cl}\,[(t_{cl}+273)^4 - (t_r+273)^4] - f_{cl}\,h_c\,(t_{cl}-t_a) - 3.05 \\
&[5.73 - 0.007(M-W) - p_a] - 0.42[(M-W) - 58.15] - \\
&0.0173M\,(5.87 - p_a) - 0.0014M(34 - t_a)\} \qquad [16]
\end{aligned}
$$

with:

$$
Fcl = f_{cl} = \frac{1.0 + 0.2I_{cl}}{1.05 + 0.1I_{cl}}
$$

$$
\begin{aligned}
t_{cl} = 35.7 - &0.0275(M-W) - R_{cl}\,\{(M-W) - 3.05[5.73 - 0.007 \\
&(M-W) - p_a] - 0.42\,[(M-W) - 58.15] - \\
&0.0173M\,(5.87 - p_a) - 0.0014M(34 - t_a)\}
\end{aligned}
$$

$$
R_{cl} = 0.155I_{cl}
$$

$$
h_c = 12.1(V)^{1/2}
$$

and:
e = Euler's number (2.718)
f_{cl} = clothing factor
h_c = convective heat transfer coefficient
I_{cl} = clothing insulation (clo. 1 clo is equal to 0.155 m²·K/W (0.88 °F·ft²·h/Btu). This corresponds to trousers, a long sleeved shirt and a jacket)
M = metabolic rate (W/m²) 115 for all scenarios

Table 4.4 The effect of adaptive behaviours on optimum comfort temperatures

Behaviour	Effect	Offset
Jumper/jacket on or off	Changes Clo by ± 0.35	± 2.2K
Tight fit/loose fit clothing	Changes Clo by ± 0.26	± 1.7K
Collar and tie on or off	Changes Clo by ± 0.13	± 0.8K
Office chair type	Changes Clo by ± 0.05	± 0.3K
Seated or walking around	Varies Met by ± 0.4	± 3.4K
Stress level	Varies Met by ± 0.3	± 2.6K
Vigour of activity	Varies Met by ± 0.1	± 0.9K
Different postures	Varies Met by ± 10%	± 0.9K
Consume cold drink	Varies Met by −0.12	+ 0.9K
Consume hot drink/food	Varies Met by +0.12	− 0.9K
Operate desk fan	Varies Vel by +2.0m/s	+ 2.8K
Operate ceiling fan	Varies Vel by +1.0m/s	+ 2.2K
Open window	Varies Vel by +0.5m/s	+ 1.1K

Source: BRE

pa = vapour pressure of air (kPa)
R_{cl} = clothing thermal insulation
t_a = air temperature (°C)
t_{cl} = surface temperature of clothing (°C)
t_r = mean radiant temperature (°C)
V = air velocity (m/s)
W = external work (assumed = 0).[9]

4.4.3 *Adaptive comfort*

Adaptive comfort models recognise that people will change their behaviour to try and stay comfortable when conditions change. They let designers increase the range of conditions that may be considered comfortable, especially in naturally ventilated buildings, so the space must have operable windows, no mechanical cooling system and the occupants must be near sedentary and have a metabolic rate between 1.0 and 1.3 met (a met is a unit equal to 58.2 W/m² (18.4 Btu/h·ft²), which is the energy produced per unit skin surface are of an average person seated at rest).

Clothing insulation values for other common ensembles or single garments can be found in ASHRAE 55.

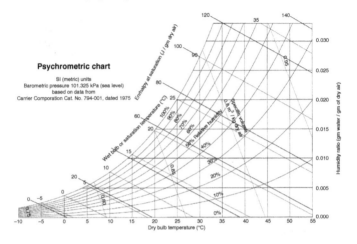

Figure 4.5 Example of a psychrometric chart

Source: WikiMedia. Creator: Arthur Ogawa

4.4.4 Psychrometric charts

Psychrometric charts can help to calculate the effect on materials and human comfort of the control of atmospheric air mixtures. A psychrometric chart is a graphical representation of the physical and thermodynamic properties of air such as dry bulb temperature, wet bulb temperature, humidity, enthalpy[10] and air density.

A psychrometric chart can be used by plotting multiple data points that represent the air conditions at a specific time on the chart. By overlaying an area that identifies the 'comfort zone' it becomes possible to use passive design strategies to extend the comfort zone. Detailed description of the application of these charts is beyond the scope of this book. A good explanation can be found at: http://bit.ly/2hToucO

4.4.5 Thermal comfort in hot humid countries and passive solar design

Extensive post-occupation evaluation of the first zero energy building 'EnerPos' constructed in the humid French tropical island of La

Reunion has shown that the users feel comfortable when the temperature is below 30°C with an air speed close to 1m/s. It is then possible to achieve thermal comfort for occupants of the classrooms and offices even during the hottest days of the year.

Nevertheless, air-conditioning is needed in the computer rooms during the six hottest weeks of the year (from the end of January to the beginning of March) due to the heat produced by the computers. Autonomy during the hours of occupation is about 80 per cent in the offices, 70 per cent in the most disadvantageous classrooms (on the first floor) and can nearly reach 100 per cent on the second floor.[11]

4.5 Degree days

To determine how much energy will be required for heating or cooling a building, and whether the required amount is actually being used, it is useful to use degree days. These have been tabulated (e.g. by meteorological services) from past weather data. A heating/cooling degree day is worked out relative to a base temperature which building occupants can tolerate without needing energy input; in the UK the predominant convention is to use 15.5°C. Elsewhere it varies, with anything between 15.5 and 18.3 °C (60 and 65°F) being used in the USA. Once chosen, however, the baseline must be consistently used for meaningful results.

Factors in the choice of baseline temperature are the other sources of heating/cooling within a building, such as body heat and heat given off by equipment, which can raise the internal temperature to normal comfort levels, generally taken to be between 18 and 20°C (70 to 75°F).

To find out the heating/cooling degree day for a given day, take the average temperature on any given day and subtract it from the base temperature. If the result is above zero, that is the number of heating degree days on that day. If the result is below or equal to zero, it is ignored.

Method: on one day the maximum temperature is 14°C and minimum is 5°C, giving an average of 8.5°C. Subtracted from 15.5°C = 7°C. A month of 30 similar days might accumulate 7 × 30 = 210 degree days. A year might add up to 2,000 degree days. The rate at which heat needs to be provided to this building corresponds to the rate at which it is being lost to the outside. This rate, for each degree of temperature difference, is the U-value (R-value in the USA and Canada) of the dwelling (as calculated by averaging the sum of the U-values of all elements) times the area of the dwelling's external surface.

To calculate cooling degree days, the degree days are noted according to the number of degrees the average outside temperature is each day above a baseline temperature that is considered comfortable, which is usually the internal temperature at which the ventilation and cooling system is switched on. Degree days are then added up when the temperature outside exceeds this baseline temperature.

4.5.1   Calculating heat loss/gain with degree days

Multiplying the rate at which a building is losing/gaining heat by the time (in hours) over which it does so reveals the amount of heat lost/gained in watt-hours (Wh) or Btu/h: this is exactly the amount of heat/cooling needed to be provided. Therefore, if the external area of the dwelling (walls, roof and floor) is A, its average U-value is U and the number of degree days is D, then the amount of heat required in kWh to cover the period in question is (SI units):

$$A \times U \times D \times 24 / 1{,}000 \tag{17}$$

where the multiplier 24 is needed to obtain the value in kWh rather than kWdays. The same method is used (without dividing by 1,000) for imperial units with the result being in Btus.

Example: if the building surface area is 500m^2, the average U-value is 3.5 and the degree days number 2,000 then the annual amount of heat required is:

$$500 \times 3.5 \times 2000 \times 24 / 1000 = 84{,}000 \text{kWh} \tag{18}$$

The above is a simplified case. The reality is more complicated. Additional factors include the thermal properties of the building, wind chill, exterior shading, heat gains within the building, the heating schedule and solar gains. Also, heat requirements are not linear with temperature and well-insulated buildings can have a lower target point.

Use of software is recommended, with the appropriate figures input. One suggestion is PHPP (see below for the procedure). There are also alternative methods of working out how much heating or cooling capacity is required – for example, the peak load requirement – determined, of course, once demand has been minimised through design (see below). More information, together with worldwide degree days for cooling and heating calculations can be found at www.degreedays.net

4.6　Breathability and humidity

Breathability is important where humidity needs to be tackled. It is not to be confused with airtightness. Breathable constructions manage humidity by being composed of hygroscopic materials, such as: wood, stone, brick, adobe, earth, straw bale, lime, hemcrete and organic sheet materials. Lime and insulating plasters are very good at absorbing and releasing unwanted moisture in problem areas. Conversely, materials which prevent air from passing through include: vinyl (incl. paint), metal, non-permeable plastic sheeting and polystyrene-type slab insulation.

In some situations, there is a danger of moisture being driven through a wall from the inside or the outside and reaching a 'dew point' where it condenses. If it fails to dry out, it can cause rot and mould. Most commonly this occurs where a non-permeable airtight barrier is on the outside of a solid wall, either combined with the weather protection (render/tile/underlay) or beneath it. Much water vapour is given off by human inhabitants in breathing, washing and cooking. This water must not be allowed to accumulate in a building.

We are learning about the vapour permeability of materials under different temperatures, pressures and humidity levels and how air infiltration into a structure, caused for example by wind, alters the picture. See section 7.7 for more information.

Notes

1　American Society of Heating, Refrigerating and Air-Conditioning Engineers, a now globally respected authority on such matters.
2　See ASHRAE Table 8.18.
3　See ASHRAE Table 8.19.
4　Harvey, L.D.D., and Siddall, M., 'Advanced glazing systems and the economics of comfort', *Green Building Magazine*, Spring, 30–5, 2008. Also found in Harvey, D., 'Reducing energy use in the buildings sector: measures, costs, and examples', *Energy Efficiency*, 2, 139–63, 2009, available to download at: http://bit.ly/2tjI8a3
5　Engineering ToolBox, available at: http://www.engineeringtoolbox.com/

6 Kishore, VVN, *Renewable Energy Engineering and Technology Principles and Practice*, Earthscan, 2009.
7 ISO = International Standards Organisation
8 ANSI/ASHRAE Standard 55 (Thermal Environmental Conditions for Human Occupancy) is a standard that provides minimum requirements for acceptable thermal indoor environments. See ashrae.org
9 Enthalpy: a thermodynamic quantity equivalent to the total heat content of a system, equal to the internal energy of the system plus the product of pressure and volume.
10 From Autodesk Sustainability Workshop: http://bit.ly/1vJDmrL
11 *Towards Net Zero Energy Buildings in Hot Climate, Part 2: Experimental Feedback*, available at: http://task40.iea-shc.org/data/sites/1/publications/DC-TP7-Garde-2011-11.pdf

5 Passive heating techniques

For heating buildings in cooler climates or seasons solar heat is absorbed by allowing sunlight to enter through glazing (preferably clean and of a high standard, see 4.2 Windows). It should fall onto dark-coloured thermally massive materials (e.g. ceramic tiles laid onto concrete/hemcrete). This heats them up during the day. During the night they release their heat into the internal space, thereby moderating extremes of temperature. Good use of thermal mass and the stack effect help to store heat and transmit it to areas not directly heated by the sun. Superinsulation and airtightness in the building help to prevent the heat from escaping.

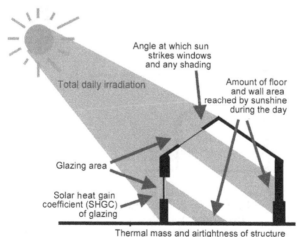

Figure 5.1 Factors affecting solar heat gain

Source: Author

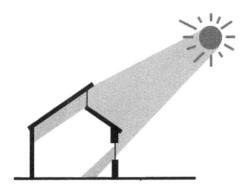

Figure 5.2 Classic passive solar heating using clerestoreys/windows/skylights in cooler climates. The sun must fall on exposed thermal mass surfaces

Source: Author

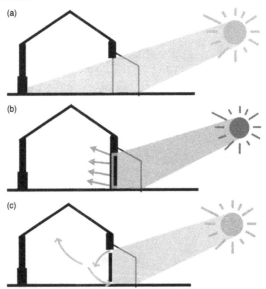

Figure 5.3 Three ways of preheating air using a conservatory or greenhouse on the side of a building facing anywhere from the south east to south west in the Northern Hemisphere or north east to north west in the Southern Hemisphere: (a): in front of glass doors or windows. (b): heating a thermally massive wall which radiates the heat indoors into the evening. (c): letting cool air circulate into the heated space through a vent at the bottom of the wall and, once heated, return through a vent at the top of the wall into the building. *Source*: Author

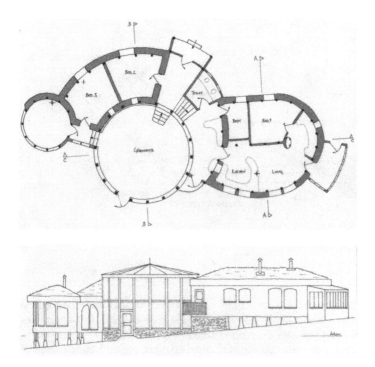

Figure 5.4 Plan and profile view of a dwelling in Wales, UK (latitude 52°N). The core of the building is the south-facing circular greenhouse/conservatory that will be tall enough to grow Mediterranean fruit trees as well as lots of other food, and capture heat that will be sent into the rest of the dwelling. The uphill, north-facing side is well insulated and contains few windows

5.1 Capturing solar energy

The floor's thermal mass should not be too thick (much over 100mm for masonry) in order to regulate the time constant for the absorption and release of the heat. A rule of thumb for the floor surface-to-glazing area is a ratio of 6:1. The floor should be insulated below and at the sides as shown in Figure 5.5. If extra heating is needed, underfloor

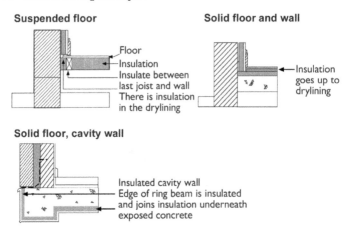

Figure 5.5 Insulation strategies for different types of floor and wall, to prevent thermal bridging

heating may be considered as the most efficient and comfortable (electric or water-based) in the coldest rooms. This is because it does not need to raise the temperature of the heat source as high as radiators for the same degree of thermal comfort. Figure 5.6 shows how it should be installed.

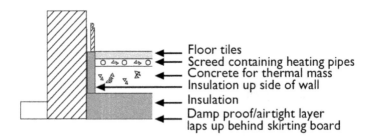

Figure 5.6 Floor with underfloor heating and insulated thermal mass

Source: Tao Wimbush

5.2   Avoiding thermal bridging

Thermal bridging conducts heat from the interior to the exterior of the building through bridging structural building components. It can cause 'cold spots' on inside walls which encourage the formation of condensation. To eliminate thermal bridges it is necessary to provide insulation all around the structure.

5.3   Transpired solar wall

A transpired solar wall is a way of trapping and directing the sun's heat into a building that is suitable for a warehouse, works unit or similar metal-skinned structure.

An outer skin of aluminium or stainless steel is coated with solar-absorbing paint. This is a few inches from the inner wall and is perforated with thousands of tiny holes. The sun heats up the metal and the hot air in the cavity is drawn up and into the building by passive means or by a fan at the top, into a heating, ventilation and air-conditioning (HVAC) system. Ducting transmits the heated air around the building. It is applicable to large industrial and commercial buildings, both new and refurbishments. Various companies

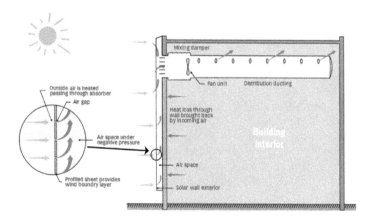

Figure 5.7 Diagram of a transpired solar wall

Source: Author

supply kit solutions. It can also be used for cooling by drawing the heat away from the interior using solar chimneys.

5.4   The stack effect

As air gets warmer it becomes less dense, more buoyant and has a tendency to rise. This effect can be used to naturally ventilate buildings. Cooler outside air is drawn into the building at a lower level, is warmed by sources of heat within the building (such as people, equipment, heating and solar gain) and rises through the building to vent out at a higher level.

Positive pressure is created at the top and negative pressure at the bottom of the building. This process can take place without mechanical assistance, simply by introducing openings at the bottom and the top of buildings.

Stack ventilation can be particularly effective as a means of naturally ventilating tall buildings that include vertical spaces which rise throughout their height – for example, buildings with central atria. These are described in section 5.6. Stack ventilation can be useful in deep buildings too, where cross-ventilation may not be sufficient to penetrate to spaces in the heart of the building.

Factors influencing the effectiveness of stack ventilation are:

* the effective area of openings;
* the height of the stack;
* the temperature difference between the bottom and the top of the stack;
* pressure differences outside the building.

The stack effect changes according to the relative temperatures and pressures inside and outside. The colder it is outside relative to inside, the more upward pressure there is. When there is hotter air outside it can push cooler indoor air down and create a need for additional mechanical cooling. Rooms adjacent to the warmer area of the stack may experience poor ventilation and unwanted heat gains.

The stack effect is helped by installing ducts from terminals situated in the ceiling of rooms that lead to terminals on the roof or in the loft space. These extract air to the outside by a combination of the natural stack effect and the pressure effects of wind passing over the roof of the building. In cases where it makes sense not to lose the heat, a heat exchanger takes the outgoing heat from the air and transfers it to incoming air that is directed to the bottom of the building (see section 5.5).

Unlike true air-conditioning, natural ventilation is ineffective at reducing the humidity of incoming air. This places a limit on the application of natural ventilation in humid climates.

Designing natural ventilation can be challenging due to the interaction between cross-ventilation and the stack effect in complex building shapes and with openings in many locations. Computer analysis is required. Calculating the stack effect is dealt with in Chapter 6.

The pressure difference (ΔP) between the outside and inside air caused by the relative difference in temperature is the driving force for the stack effect and it can be calculated with the equations presented below. For buildings with one or two floors h is the height of the building. For multi-floor, high-rise buildings h is the distance from the openings at the neutral pressure level (NPL) of the building (the location where the interior and exterior pressures are equal) to either the topmost openings or the lowest openings. It can be calculated from the following equation:

$$\Delta P = Cah \left(\frac{1}{T_o} - \frac{1}{T_i} \right)$$

where:
ΔP = available pressure difference, in Pa (psi in imperial)
C = 0.0342 (0.0188 for imperial units)
a = atmospheric pressure, in Pa (psi in imperial)
h = height or distance, in m (ft in imperial)
T_o = absolute outside temperature, in K (°R in imperial)
T_i = absolute inside temperature, in K (°R in imperial).

5.5 Double-skin façades

Double-skin façades are used for high-rise buildings. They consist of two glass skins separated by air, which acts as an insulating barrier against temperature extremes, noise and wind. The outer layer is like a standard curtain wall. The main layer is commonly insulating. Between them, shading devices are often located.

Pros:

- can help to moderate the internal climate with insulation, cooling and shading;
- passive design with user control;
- can help with acoustic protection from street noise;
- can protect shading devices from rain and wind;
- can allow for safe natural ventilation and passive night cooling;
- can renovate the building skin without perturbing the occupants;
- can protect the covered exterior wall.

Cons:

- expensive (two to four times);
- increased ecological footprint;
- similar energy savings may be achieved with conventional high-performance, low-e windows;
- decrease in usable floor space;
- potential problems with condensation, dirt;
- building energy modelling is harder due to varying heat transfer properties within the cavity;
- life-cycle costing must be taken into account, including ease of maintenance.

The double façade is typically situated on the sun-facing side of a building. Closer to the equator it may be the east and/or west. If there is a strong prevailing wind, this may also drive cooling using the façade, so it can be situated according to the windward/leeward sides. In both cases solar and wind shadows created by nearby buildings or geographic features will contribute to a determination of the design, perhaps even ruling out the use of the double façade altogether.

Four basic types may be defined:

1. buffer
2. extract-air
3. twin-face
4. hybrid.

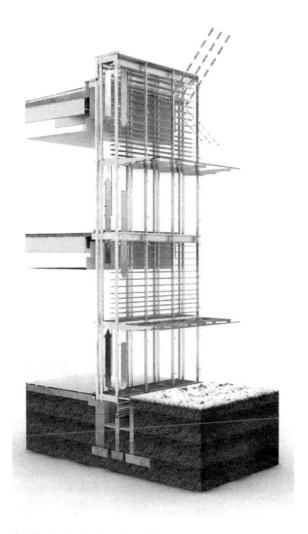

Figure 5.8 Cambridge Public Library double-skin façade with shading

Source: Hyun Tuk on progressivetimes.files.wordpress.com, uploaded under public license

Buffer:

- permits daylight entry while increasing heat and acoustic insulation;
- uses two layers of single glazing spaced 250 to 900mm apart, sealed, permitting fresh air in through further controlled means – either a separate HVAC system or box-type windows within the overall double skin;
- shading devices can be included in the cavity.

Extract-air:

- contains a second single glazing layer on the interior of an outer insulating façade of double-glazing;
- cavity is 150 to 900mm, dependent upon requirements for cleaning and shading devices;
- natural ventilation is not employed, so suitable for locations with high noise, wind or fumes;
- the air between becomes part of the HVAC system and when heated is extracted through the cavity with the help of fans;
- fresh air is supplied by HVAC;
- does not reduce energy use;
- occupants may not adjust the temperature;
- no longer favoured.

Twin-face:

- a conventional curtain wall or thermal mass wall inside a single-glazed building skin;
- cavity is >500mm to permit cleaning;
- simpler, with less energy required than extract-air type;
- includes openings to permit natural ventilation;
- shading devices may be included;
- the internal wall is a thermal buffer/heat store/insulant;
- the glazed skin offers wind protection and permits noise- or turbulence-free interior openings for fresh air;
- night-time cooling of the interior is permitted.

Hybrid:

- combines one or more of the above.

Cavities may be subdivided and compartmentalised, or not, with the following features.

Undivided cavities:

- permit use of stack effect;
- base openings admit cool air;
- top openings can vent hot air;
- can be transformed into atria and include plants (oxygen, cooling and shading);
- usually deeper;
- risk of stack effect rapidly spreading any fire;
- can transmit noise, dirt and odours.

Divided cavities:

- different configurations possible;
- stack effect may not be used for air movement;
- yield fire protection and sound insulation;
- usually narrower;
- each compartment contains intake and outward openings.

Four double façade types using cavities may be described:

1. *Box window*

- divided horizontally and vertically often coincident with room sizes;
- each requires own air intake and extract openings in each layer.

2. *Shaft-box*

- box windows with continuous vertical shafts extending over several storeys;
- permits use of stack effect;
- alternates boxes and shafts;
- requires fewer openings on the external skin (less noise and other infiltration);
- appropriate for lower-rise buildings.

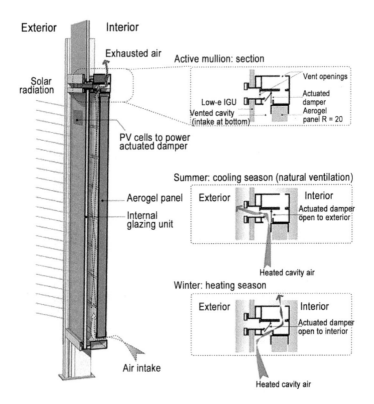

Figure 5.9 Divided cavity double-skin façade with actuated damper

Source: Hyun Tuk on progressivetimes.files.wordpress.com, uploaded under public license

3. *Corridor*

- cavity corresponds to length and height of floors;
- designed for acoustic, fire-protection or ventilation reasons;

- deployed at the corners of buildings to prevent cross-drafts;
- air intake at floor, extract at ceiling;
- stack effect may not be used;
- divisions are usually made where there is a pressure difference along the façade.

4.   *Multi-storey*

- divided horizontally and vertically coincident with multiple rooms;
- may extend around the entire building without divisions;
- air intake is at the base, exhaust at the top;
- does not necessarily require openings all over the exterior of the façade.

Considerations for cavity systems:

- better control and access to daylight is possible;
- higher degree of interior comfort for occupants;
- steps must be taken to limit any fire hazard within the cavity;
- potentially higher level of user control of façade system;
- access to natural ventilation;
- higher upfront cost may be mitigated by long-term savings on energy use and occupants' well-being;
- condensation in cold climates may occur on surfaces within the cavity. Ventilated cavities remove humid air from the interior, either to the exterior or to the interior HVAC system return air, depending on the system;
- if condensation is on cold outer pane, can be mitigated by double-glazing or use of insulated glass;
- sunshades in the cavity alter the heat levels and ventilation paths. Both sides of the cavity must be ventilated.

5.6   Atria

Atria are useful for introducing solar light and solar heat into areas of a building that are not directly heated by the sun, particularly if they occupy a central well, with other rooms coming off. They are also used for providing ventilation.

Ventilation is provided because an atrium or solar chimney or similar vertically spanning space can act as an exhaust stack. Filled with buoyant air it is a source of driving pressure. The taller the atrium, the greater the temperature difference and stack effect. Atria need to be managed correctly. Some considerations are:

- if the atrium vent is too small compared to the storey vents, reversed flows through the top storey produce an uncomfortable, stuffy internal environment;
- if there is incorrect relative sizing of storey vents, this results in insufficient ventilation and overheating on upper storeys;
- if the atrium vent is too large, then the exchange flow at this vent leads to reduced flow rates through the storeys below;
- exchange flows in which simultaneous inflow and outflow occur at a high-level vent can also reduce net ventilation flow rates and contravene fire regulations because an inflow occurs at what should be an outlet vent.

Atria in buildings therefore need careful design and modelling using multizone software. An effective design depends largely on the correct relative sizing of vents. A successful example is Manitoba Hydro Place, in Winnipeg, Canada. This 22-storey building makes use of three atria and a solar chimney for passive supply and extract of air. The building's form, a sophisticated BMS design, and over 25,000 sensors contribute to its success.

5.6.1   Recommendations for a successful approach

Choose a simple design. This is easier to communicate to the client and team, and to work with.

Include as core design parameters: vent sizes, heat inputs, target air temperatures and building geometry. The design stages would be:

1. to focus only on buoyancy-driven ventilation, designing for a worst-case scenario where no wind is available to assist ventilation;

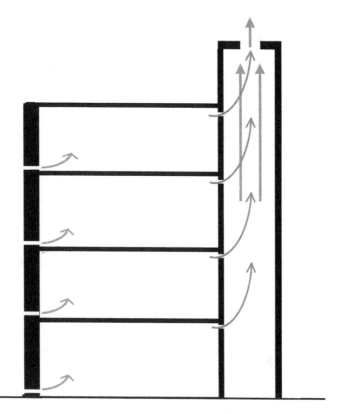

Figure 5.10 Wind flow as a result of successful integration of an atrium with a building: prevailing wind arrives from the left. Forward flow operates on all storeys, with fresh air supplied at floor level from the windward side. Warm air is vented to the atrium at ceiling level. Vents are controllable but their maximum size is related to the average number of people expected to occupy each level. The atrium is taller than the building to allow the stack effect to ventilate the top storey. This building contains four storeys but the principle works at other heights

Source: Author

2. to specify on a per-person basis with vent sizes and heat gains;
3. to compare the theoretical flow rate per person through the top storey with, and without, the addition of an atrium. The top storey would be expected to be the worst-performing storey in terms of passive ventilation;
4. to design the integration of the atrium to perform optimally with this in mind;
5. to then proceed with optimising the performance of the remainder of the building by the development of some design principles.

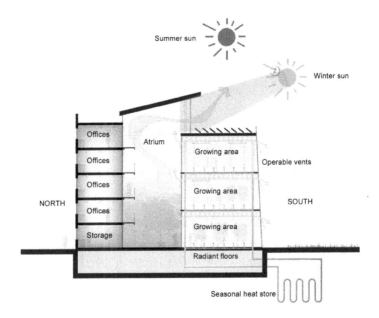

Figure 5.11 An application of the atrium in an indoor farm in Newark, New Jersey, USA (latitude 41°N). Light can illuminate the rooms on the north, cold side in the winter months

Source: Author

These successful design principles would include:

- the vent size per person to increase on successively higher floors to compensate for the reduction in driving stack pressure;
- an equal flow rate per person and equal air temperature on all storeys;
- equal per-person vent sizes at the atrium outlet and in the top storey;
- the atrium should extend, if possible, at least one storey height above the top storey to ensure an enhanced flow through all storeys. If this is not possible, the top storey to be disconnected from the atrium with its ventilation provided by a separate system;
- fresh air to be supplied at floor level, with warm exhaust to the atrium exwpelled at ceiling level.

See Figure 6.20 in the next chapter for another example.

5.7 Mechanical ventilation with heat recovery (MVHR)

Air-to-air heat exchangers, where used, supplement passive heating systems in situations where they alone cannot complete the job. Used in many Passivhaus projects, they take the heat from the outgoing air and transfer it to the incoming air. Termed 'mechanical ventilation with heat recovery' (MVHR), they employ fans to direct the air through the baffles which exchange the heat, since the stack effect is not usually sufficiently strong to be reliable for this purpose. These units can take up considerable space. There is a potential cost, maintenance and energy-use trade-off between using MHVR and prioritising energy efficiency with controlled natural ventilation. Finding a balance requires sensitive modelling and adjustments at the design stage. It may be more efficient to use pre-heated (in winter) or pre-cooled (in summer) air from a ground source heat pump and use natural ventilation techniques as outlined in the following section. Different means of introducing air may be used in summer and winter.

Where MHVR is used in a Passivhaus, ventilation units must have been certified by the Passivhaus Institute and possess a heat recovery efficiency of ≥ 75 per cent, with a specific fan power of ≤ 0.45 Wh/m^3. The fans must be quiet, efficient, variable speed and easy to maintain. They come with installation instructions to restrict noise pollution.

6 Passive cooling techniques

Cooling is needed in many different climates, from those with seasonal extremes to those with hot, dry climates and hot, humid climates. Passive cooling techniques can help reduce demand for fossil fuels in all climates, but each type of climate demands a different approach. The best way to prevent overheating and unwanted glare from direct sunlight is to shade the sunlight from reaching the building or to prevent solar gain in the interior of the building during the periods when it is in danger of overheating. Once hot air is inside the building, or if the unwanted temperatures are generated within the building, passive techniques should be encouraged to remove it.

'Passive', here, relates to the dominant design and functioning of the building. In practice, a number of active devices and methods are likely to be used to add to comfort and performance in some areas of the building.

The cooling demand may be calculated using the same methods as for heating demand, as outlined in Chapter 4.

6.1 Hot, humid climates

Shading plus an open-building approach work well in warm and humid locations where the temperature is tolerable and does not change much from day to night. Daytime cross-ventilation is used to create an internal air flow that helps to maintain comfort levels. Avoid the use of pools and areas of open water as these encourage insects and humidity.

In locations and situations where this is not appropriate, two possible strategies are available:

- an airtight building with mechanical cooling and dehumidification, which can be driven by renewable energy; or
- Passivhaus, with careful attention to vapour control (see section 7.7 for more information).

6.2   Hot, dry climates

6.2.1   *Techniques*

Below is a list of techniques available for hot, dry climates with a large temperature variation from day to night use. Many of these are discussed in detail in this chapter:

- large interior thermal mass;
- exterior superinsulation, especially the roof;
- shading (e.g. overhanging roofs, shutters);
- reflective external coating on the roof and sun-facing side – can reduce up to 10 per cent of the total cooling load – the most cost-effective technology;
- OR 'green' (with vegetation) roof;
- insulation, beneath roof deck;
- a radiant barrier beneath the roof deck;
- location-appropriate exterior shading above windows;
- natural ventilation;
- daytime cross-ventilation if it is required to maintain indoor temperatures close to outdoor temperatures;
- avoidance of drawing in unconditioned replacement air that is hotter or more humid than interior air;
- night-time ventilation;
- closing windows and shutters in the early morning to keep out the hot daytime air;
- use of phase change materials;
- air vents (controllable for ventilation);
- diode roofs;
- roof ponds;
- wind catchers/towers;
- domes and curved roofs;
- fresh air brought in through a dehumidifier through the crawl-space or basement through underground pipes;

- solar-powered absorption chillers;
- direct evaporative cooling.

Cooling is provided by radiant exchange with massive walls and floor, plus optional techniques detailed below. In practice, large openable windows, probably with slatted shutters, are common, so air movement allows personal cooling, especially by evaporation. As windows opened at night-time make for easy access by burglars, anti-burglar bars are fitted. Open staircases, etc. may provide stack effect ventilation, but observe all fire and smoke precautions for enclosed stairways.

Curved roofs and air vents are often used in combination in places where dust-laden winds make wind towers and open windows impracticable. A hole in the apex of a domed or cylindrical roof with a protective cap over the vent directs the wind across it and provides an escape path for hot air collected at the top. Arrangements may be made to draw air from the coolest part of the structure as replacement, to set up a continuous circulation and cool the spaces.

Large buildings require detailed modelling. Power may be required for anti-stratification fans and ducts.

6.2.2 Design guidelines

The specific approach and design of natural ventilation systems will vary based on building type and local climate. However, the amount of ventilation depends critically on the careful design of internal spaces, and the size and placement of openings in the building. Some design recommendations are listed below.

- Maximise wind-induced ventilation by siting the ridge (with ridge openings) of a building perpendicular to the summer winds. Approximate wind directions are summarised in seasonal 'wind rose' diagrams available from the National Oceanographic and Atmospheric Administration (NOAA). However, these are usually based on data taken at airports; actual values at a remote building site can differ dramatically; therefore, if possible, research should be done at the site.
- Ridge vents should be provided. These are openings at the highest point in the roof that offers a good outlet for both buoyancy and wind-induced ventilation. They should be free of obstructions to allow air to freely flow out of the building.

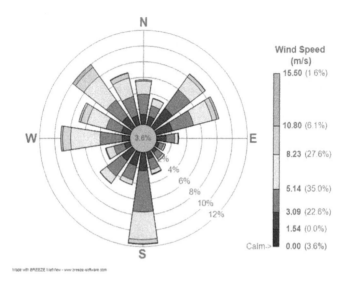

Figure 6.1 Example of a wind rose. This one is for LaGuardia Airport, New York. Created by Breeze Software

- Buildings should, if possible, be sited where summer wind obstructions are minimal.
- Naturally ventilated buildings should be narrow. It is difficult to distribute fresh air to all portions of a building much wider than 14 metres (45 ft) using natural ventilation.
- In wider buildings an articulated floor plan is often adopted. This involves giving each room two separate supply and exhaust openings, with the inlets as low as possible and the outlets as high as possible in the walls in order to make the most use of the stack effect. They would also be on opposite sides of the room and offset from each other to encourage mixing, and there should be a minimal number of obstructions to air flow.
- Window openings should be operable by the occupants.
- Allow for adequate internal air flow between the rooms of the building. When possible, interior doors should be designed to be open to encourage whole-building ventilation. If privacy or fire protection is required, ventilation can be provided through high

louvres or transoms, and areas of the building divided into zones which are modelled separately.

- Consider the use of clerestoreys or vented skylights for stale air to escape in a buoyancy ventilation strategy. The light well of the skylight might act as a solar chimney to augment the flow. Openings lower in the structure, such as basement windows, must be provided.
- Open staircases provide stack effect ventilation, but observe all fire and smoke precautions for enclosed stairways.
- In buildings with attics, ventilating the attic space greatly reduces heat transfer to conditioned rooms below. Ventilated attics are about 17°C (30°F) cooler than unventilated attics.
- Ceiling and whole-building fans can provide up to 5°C (9°F) effective temperature drop at one tenth the electrical energy consumption of mechanical air-conditioning systems.
- Determine if the building will benefit from an open- or closed-building ventilation approach. A closed-building approach works well in hot, dry climates where there is a large variation in temperature from day to night. A massive building is ventilated at night, then closed in the morning to keep out the hot daytime air. Occupants are then cooled by radiant exchange with the massive walls and floor.

6.3 Shading tactics

Shading is a principal tactic to protect the building from direct sunlight. Some methods for providing shading are:

- covering of roofs and courtyards with deciduous vegetation (allowing winter solar access) such as creepers or grapevines permits evaporation from the leaf surfaces to reduce the temperature. During the day, this temperature may be less than the sky temperature;
- horizontal overhangs or vertical fins prevent overheating while preserving natural daylighting;
- for east and west walls and windows in summer: vertical shading and/or deciduous trees and shrubs;
- for south-facing windows: horizontal shading;
- shutters, closed in the day;
- highly textured walls leave a portion of their surface in shadows cast by the protruding areas so absorb less heat;

- green roofs, earthenware pots filled with water laid out on the roof and highly reflective surfaces (e.g. painted with titanium oxide white paint/whitewash) are all techniques practised widely. All of these prevent heat from penetrating the roof, the first two mostly by absorbing some of it, the latter by reflecting it;
- coverings that in the daytime insulate the roof but automatically withdraw at night exposing the roof to the night sky, allowing heat to leave by radiation and convection.

6.3.1 Shading devices

Shading must obstruct unwanted solar heat gains from reaching the building skin and interiors but not interfere with desired winter-time solar gains. It must also control intense daylight by diffusing it and admitting sufficient for daylighting purposes while guaranteeing an unobstructed view from windows and permitting required amounts of controlled ventilation.

Shading may be external or internal, the former being more effective since it prevents unwanted heat entering the building. The amount of overhang or shading should be calculated according to the sun's behaviour throughout the day and seasons. Modelling software is often used, along with sun path diagrams or sunpegs. See section 2.6 and Figure 6.2.

The following is a procedure for designing shading devices:

1. Determine the times when shading is needed to prevent overheating or glare, typically any time when outdoor air temperature exceeds 70°F (21°C) at latitude around 40°. For every 5° latitude change towards the Equator increase this limiting temperature by 0.75°F (0.2°C).
2. Use a sun path diagram (see section 2.6) to determine the position of the sun when shading is required. Mark the overheated period on the sun path by constructing a table of average temperature for every hour in every month. These are the times of the overheating period.
3. Transfer the boundary lines of the overheated period to the sun path diagram.
4. Determine the type and position of the shading device by plotting on a protractor with the same scale as the sun path diagram the shading mask of a given shading device.

5. Determine the dimensions of the shading device so that it does permit in light and heat during the time is required.

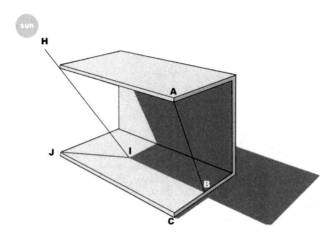

Figure 6.2 Diagram showing the angles used when calculating shade from overhangs: angle of incidence – the angle between the sun's rays and the face of the building; true altitude (HIJ) – the angle between the sun's rays and the horizon; profile angle (ABC) – the angle between Normal to window and the rays of the sun perpendicular on the window plane. (Normal is a line perpendicular to the plane of the window.) Angle HA is the horizontal plane and angle IB is the horizontal axis in the plane of the window

Source: Author

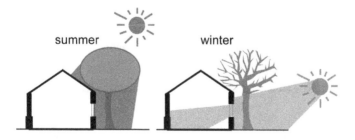

Figure 6.3 The use of deciduous trees to provide shade in summer and admit light in winter

Source: Author

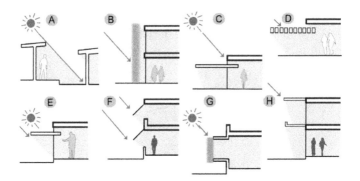

Figure 6.4 Different shading types. A: exterior structure with overhang; B: exterior obstruction; C: exterior shading overhang extended inside through glazing (NB: break thermal bridges); D: C but with vertical fins; E: user-operated swinging shutter; F: adjustable shutters; G: extended window bay with semi-opaque or horizontal-finned glazing; H: overhang for each storey

Source: Author

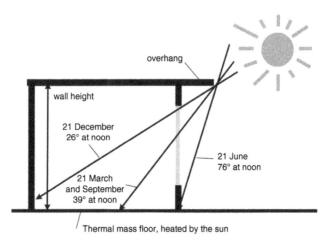

Figure 6.5 The outcome of the successful design of overhangs relative to latitude, permitting the low sun in winter to heat the interior, but excluding it during the summer

Source: Author

6.4 Natural ventilation

The design of natural ventilation systems varies, depending on building type and local climate. The amount of ventilation depends on the careful design of internal zones, and the size and placing of openings.

The total air flow due to natural ventilation results from the combined pressure effects of wind, buoyancy caused by temperature and humidity (the stack effect, see section 6.5), plus any other effects from sources such as fans. The air flow from each source can be combined, as discussed in the ASHRAE Handbook (see Chapter 10).

Pros:

- suitable for many types of buildings in mild or moderate climates;
- the option to be able to open windows is often popular, especially in pleasant locations and mild climates;
- usually cheaper to install and run than mechanical systems;
- high air flow rates for cooling and purging are possible, given many openings;
- short periods of discomfort during periods of warm weather can usually be tolerated;
- no mechanical plant room space is needed;
- minimum maintenance;
- no fan or system noise.

Cons:

- inadequate control over ventilation rate could lead to poor indoor air quality and excessive heat loss;
- air flow rates and patterns are not constant;
- fresh air distribution in large, deep-plan and multi-roomed buildings may be difficult;
- high heat gains may necessitate use of mechanical cooling and air handling instead of natural ventilation;
- unsuitable for noisy and polluted locations;
- possible security risk;
- heat recovery from exhaust air is not generally practicable;
- may not be suitable for severe climates;
- occupants must normally adjust openings to suit demand;

- filtration or cleaning of incoming air not usually possible;
- ducted systems require large-diameter ducts and restrictions on routing.

6.4.1 Wind pressures

Wind pressures depend on speed and direction, the location and surrounding environment of the building, and its shape. Wind pressures are generally high on the windward side of a building and low on the leeward side; they will also be lower on the roof and the sides when the wind is blowing close to horizontal. Pressure on building surfaces may be expressed as:

$$P_w - P_o = C_p \left[\frac{\rho \cdot v_w^2}{2} \right]$$

[1]

where:
P_w = mean pressure on the building surface (N/m^2 or Pa)
P_o = static pressure in undistributed wind (N/m^2 or Pa)
v_w = mean wind velocity (m/s)
ρ = density of air (kg/m^3)
C_p = surface pressure coefficient.

For stand-alone buildings of simple form, or ones much higher than surrounding buildings and free of obstruction, British Standard BS5925 gives average surface pressure coefficients.

To equalise pressure, fresh air will enter any windward opening and be exhausted from any leeward opening. In seasonal climates, in summer, wind is used to supply as much fresh air as possible while, in winter ventilation is normally reduced to levels sufficient to remove excess moisture and pollutants. An expression for the volume of air flow induced by wind is:

$$Q_{wind} = K \times A \times V$$

[2]

where:
Q_{wind} = volume of air flow (m^3/h)
   A = area of smaller opening (m^2)

V = outdoor wind speed (m/h)
K = coefficient of effectiveness.

The coefficient of effectiveness depends on the angle of the wind and the relative size of entry and exit openings. It ranges from about 0.4 for wind hitting an opening at a 45° angle of incidence to 0.8 for wind hitting directly at a 90° angle.

For a building with many partitions and openings, the relative sizes of the openings and the wind direction determine the relative pressures and therefore the behaviour of air flows. The provision of large openings on the windward side puts the building under positive pressure and allows it to cool. Reversing this design will cause it to warm.

6.4.2 Buoyancy

Buoyancy results from a difference in air density, which depends on temperature and humidity (cool air is heavier than warm air at the same humidity and dry air is heavier than humid air at the same temperature). This effect can be used to promote ventilation.

The two available techniques are temperature-induced (stack ventilation) or humidity induced (using a cool tower). The two can be combined.

6.5   Stack effect

As discussed in Chapter 4, the 'stack effect' is used to help circulate heat or introduce cooler air in a building and moderate its internal temperature and climate. Warmer air will rise to exit at the top through open windows, vents, chimneys or leaks. This causes reduced pressure lower in the building, drawing in colder air through any openings.

The effect is proportionate to the height difference between openings and the relative size of the openings (using the Venturi effect, see below). In large or high-rise buildings with a well-sealed envelope, pressure differences can become acute, causing uncomfortable air speeds, and should be managed with the help of zoning (having areas sealed from each other), mechanical ventilation, partitions and floors, with fire doors to prevent the spread of fire.

The flow rate induced can be calculated from:[1]

$$Q = C \cdot A \cdot \sqrt{2gh\frac{T_i - T_o}{T_i}}$$

[3]

where:

Q = stack effect draft (draught in British spelling) flow rate, m³/s or ft³/s

A = flow area, m² or ft²

C = discharge coefficient

g = gravitational acceleration, 9.81 m/s² or 32.17 ft/s²

h = height or distance, m or ft

T_i = average inside temperature, K or °R

T_o = outside air temperature, K or °R.

This equation assumes that the resistance to the draft flow is similar to the resistance of flow through an orifice characterised by a discharge coefficient C – which would need to be determined (in the absence of this, a rough figure of 0.7 can be used). The pressure patterns for actual buildings continually change with the relative magnitude of thermal and wind forces acting together, which determine the total pressure difference across the building envelope, as shown in Figure 6.6.

Stack effect ventilation is effective in winter or whenever the indoor/outdoor temperature difference is large (e.g. at night when it may be cooler for ventilating heat absorbed during the day in thermal mass or phase change materials). Conversely, it will not work in summer or when the ratio is low or inverted. In these cases a solar chimney may be brought into play, using thermostat-operated vents (see below).

In a multi-storey building, the rooms or zones on the outside faces may be separately controlled, depending upon the degree of sophistication present in the building, its size and location. If hot air is present above a set temperature, it is allowed to escape at whatever rate is necessary to preserve the comfort of this zone's occupants, via controlled venting into a vertical space – atrium or stairwells/lift shafts – positioned centrally or on corners (perhaps with glass sides to aid the process).

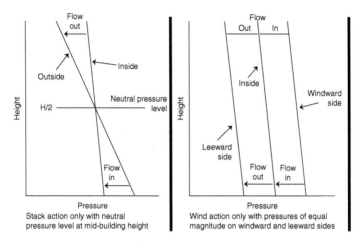

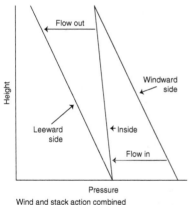

Figure 6.6 Pressure variation with height in a building due to the stack effect, wind and the combination

At the top of this space, louvres, perhaps in clerestoreys or skylights, allow a controlled amount of hot air to escape, again, at whatever rate is necessary for comfort of the whole building's occupants. Basement windows allow cool air in. If present, for most of the time the air-conditioning should need no mechanical input.

Strategies:

- offset inlet and outlet windows across the room or building from each other;
- make window openings operable by the occupants but controlled by the building management system for security;
- provide ridge vents at the highest point in the roof that offers a good outlet for both buoyancy and wind-induced ventilation;
- allow for adequate internal air flow;
- in buildings with attics, ventilate the attic space to reduce heat transfer to conditioned rooms below.

Slots in doors permit air to travel from one room to another within a zone. Duct sizes for natural ventilation should be a minimum of 150cm^2 in profile to permit sufficient air flow. Long, horizontal ducts should be avoided. Easy maintenance should be possible.

It is not always possible for a building to be completely naturally ventilated. In such cases fan assistance is required (an 'active' intervention). Thermostats, dampers and fans would be connected to a building energy management system. Fans (motors) should have variable speed drives for efficiency.

Air intakes are sited in places where they avoid draughts. Air intakes with built-in thermostats are available that regulate input according to outdoor and indoor temperature. The heating of intake air can also be provided using a solar collector installed beneath a window. Here, fresh air enters through a gap between the wall's insulation and exterior black sheet metal covered with glass. The intake must be adjustable to prevent overheating in warm weather. Perforated walls are used in industrial buildings (see section 5.3).

Evaporative cooling in dry climates (see below for more information) can be achieved by letting incoming air filter through expanded clay pellets that are sprinkled with water in the summer. The air is cooled both by evaporation and by the underground temperature, if combined with a buried pipe from an outside intake some distance from the building.

Pros:

- provides good winter driving force in cold climates;
- can relieve problem of single-sided ventilation by providing stacks in the interior of the building;

- can be used together with wind-induced ventilation by locating the roof termination in the negative pressure region generated by the wind.

Cons:

- each room should be individually ducted since shared ducts may result in cross-contamination between zones;
- potential for reverse flow (downdraught) if the column of air in the stack becomes cold;
- requires temperature differential between inside and outside.

6.6 Cross-ventilation

Single-sided ventilation is popular because openings are located on one face only, but there is no defined exit route for air and so poor ventilation may result. The maximum depth of fresh air penetration may be 2.5 times the ceiling height. Therefore, single-sided natural ventilation should be avoided. By contrast, cross-ventilation uses a flow of air through openings, such as doors, windows or grilles, on opposite sides of a room or building and is much more effective. Cross-flow designs form the basis of best practice in natural and mixed-mode ventilation systems. To succeed, openings must be well distributed and flow paths within the building available and modelled. Wind pressure can also drive single-sided and vertical ventilation.

Pros:

- can be adopted whenever a building is exposed to the prevailing wind;
- 'open' air flow path presents minimum resistance to air flow, providing good ventilation;
- for equivalent size of openings, will provide more reliable ventilation than single-sided;
- good for small, or single-ownership/occupancy, or open-plan buildings.

Cons:

- may not be useful where there is noise and pollution, in dense urban areas, close to the ground or in heavily vegetated sites, where the prevailing wind may be of too low a speed;

- flow paths may conflict with acoustic separation between internal spaces; bypass ducts can help reduce this;
- fire risk must be dealt with by dampers and automatic fire doors;
- not (in general) suitable where floor widths are greater than five times the floor–ceiling height in single-storey buildings;
- cross-flow of 'used' air into further occupied spaces should be avoided;
- not suitable for multi-occupancy buildings (use double-skin façades, see section 5.5).

In general, wind speed and direction is variable. Wind generates complex pressure distributions on buildings, particularly in urban environments. In open sites, regional wind data can be consulted. In urban environments, provide as many controllable openings as possible, well distributed over the building envelope. Openings must be controllable to cover the wide range of required ventilation rates and the wide range of wind speeds. The more the opening areas are distributed, the more likely there will be a pressure difference between openings to drive the flow – that is, many small openings are better than one large opening, as in the examples in Figure 6.7.

If the windward or leeward side of the room/building may be partitioned to form a corridor, a Venturi effect may be encouraged. A bypass duct would pressurise the corridor with fresh air and allow independent control to the occupant of the leeward room. The duct could be within a ceiling zone, or a useable annex to the circulation space, between two windward rooms.

6.6.1 Deep open-plan layouts

The main limitation will be providing sufficient fresh air via the windward openings to meet demand without causing disturbance to occupants close to the openings. Air quality will also diminish as it crosses the space. The stack effect must be utilised in addition, using layouts like those on the bottom row of Figure 6.7, either:

1. with an air intake via an atrium and the outlet via stacks or windows on the perimeter; or
2. with the stack (e.g. an atrium or stairwell/liftshaft) drawing air through openings in the perimeter.

The following rule of thumb may be used to assess the potential for single-sided and cross-ventilation. The depth of plan over which

Cross-ventilation plans

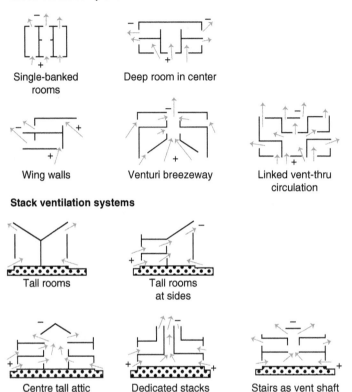

Single-banked
rooms

Deep room in center

Wing walls

Venturi breezeway

Linked vent-thru
circulation

Stack ventilation systems

Tall rooms

Tall rooms
at sides

Centre tall attic

Dedicated stacks

Stairs as vent shaft

Figure 6.7 Cross-ventilation and stack ventilation strategies

Table 6.1 The depth of plan over which ventilation can be expected to work
is specified as a ratio of the floor to ceiling height

Ventilation configuration	Depth of floor/ceiling height
single-sided, single opening	1.5
single-sided, multiple opening	2.5
cross-ventilation	5

Table 6.2 Area of openings required to use natural ventilation dependent upon heat gains to be removed

Heat gains	Total area of openings as % of floor area
< 15 W/m^2	10
15 – 30 W/m^2	20
> 30W/m^2	25

Source: RIBA

ventilation can be expected to work is specified as a ratio of the floor to ceiling height.

The greater the heat gains the more ventilation is required, so gains should be reduced by other means first.

Since over-ventilation will increase heat loss and nullify efforts to make the envelope airtight, fit slot ventilators. These are self-balancing: as wind pressure increases, they automatically close up.

Openings can be under intelligent control, with the control parameter being temperature and/or CO_2 concentration.

6.7   Atrium ventilation

The above principles are often combined with the use of an atrium than runs the full height of the building. See section 5.6 for more detail.

Pros:

• provides an extract driving force on the core of the building to drive cross-flow ventilation through surrounding rooms;
• the zone above the occupied area can trap waste heat which can add to the stack driving force.

Cons:

• flow can be upset by wind forces.

6.8   Bernoulli's principle and the Venturi effect

Bernoulli's principle multiplies the effectiveness of wind ventilation and is an improvement upon simple stack ventilation. However, it needs wind, whereas stack ventilation does not. In many cases, designing for one effectively designs for both, but some strategies can be employed to emphasise one or the other. For example, a ventilation chimney optimises for the stack effect, and wind scoops optimise for Bernoulli's principle.

The principle uses wind speed differences to move air, based on the fact that the faster air moves, the lower its pressure. Outdoor air farther from the ground is less obstructed, with a higher speed, and thus lower pressure. This can help suck fresh air through the building. The success of this process depends upon the building's surroundings: there must be no obstructions to wind flow.

The BedZED development in south London (Figure 6.8) utilises specially designed wind cowls which have both intakes and (larger)

Figure 6.8 BedZED development, London (latitude 51.5°N), showing use of wind cowls for ventilation

Source: Bioregional

Figure 6.9 A typical Iranian badgir in Meybod, Province of Yazd, Iran (latitude 32.2°N). This traditional cooling system is made with mud bricks and adobe and uses the air circulation between two towers passing through a dome refreshed by the flow of water into an underground channel named Qanat

Source: Patrick, C., or Dynamosquito under CC BY-SA 2.0

outlets; fast rooftop winds get scooped into the buildings. The larger outlets create lower pressures to naturally suck air out.

6.8.1 Ventilation chimneys

Ventilation chimneys include caps to prevent backdrafts caused by wind. These adjust according to the wind intensity and direction and increase the Venturi effect. Turbines may be deployed to increase ventilation. Self-regulating turbine models are available. There are several styles of passive roof vents – for example, open stack, turbine, gable and ridge vents – which utilise wind blowing over the roof to create a Venturi effect that intensifies natural ventilation.

6.9 Wind towers

Wind catcher towers (*badgir*) in the Middle East have long been a staple feature of traditional architecture. Their design is now being studied and the learning applied to new forms of architecture, such as the design of atria (see section 5.6).

Wind towers, catchers or chimneys enhance Bernoulli's principle. A tower is located on the side of the building to be ventilated. It has openings in the side at the top, with an opening designed to face, or to turn to face, using a vane, the prevailing wind. The most prevalent type of cool towers induce air into the top of the tower, then pass it over a wet medium, causing it to cool. It then falls to the base of the building and is drawn in through passive stack ventilation either caused by the building's design, or by the addition of a solar tower (see below). It will work without wind, but wind will improve the air flow.

Generally, cool towers without fans stand from 5–10 metres (20–30 feet) proud of the roof and are 2–3 metres (6–10 feet) square. Their effect can be helped by the addition of a solar-powered fan to drive the cool air downwards. Inside, it may be separated into two or more shafts, which allow air to move easily up and down the tower at the same time.

1. In the daytime ambient air blows into the openings drawn down by pressure differences to the base, and sometimes even underground in a basement, where it cools. It is then allowed to

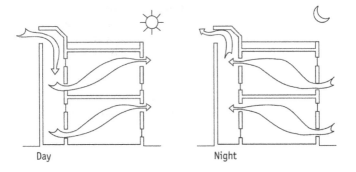

Day   Night

Figure 6.10 The principle of a wind tower

Source: Author

circulate upwards through the building and exit through openings near the top.

2. At night-time, there is a reversal of air flow: cooler air enters the bottom of the tower after passing through the rooms. It is heated up by the warm surface of the wind tower and leaves in the reverse direction, sucking warm air out of the building. In this way, the heat in the thermal mass of the building, gained during the day, is allowed to escape.

Fabric umbrellas or flaps on the tops of the towers may act as large dampers, lifting to exhaust hot air and closing on cool days to conserve heat. In some buildings the flaps have vanes to turn with the wind, so the wind may enter the chimneys and be drawn down, or it may suck unwanted hot air out, depending on which way the vent faces. This is only effective where outdoor humidity is low.

Pros:

*   air is drawn in at high level, where pollutant concentration is usually lower than at street level;
*   can be integrated with a mixed mode fan to ensure reliable operation under low wind speed conditions;
*   possible to supply air into deep-plan spaces.

Cons:

*   reliable wind force is required unless combined with mixed mode;
*   can usually only provide fresh air to up to three-storey buildings;
*   possible conflict with stack-driven ventilation;
*   cold draughts are possible in winter periods.

6.9.1   Measuring flow rate

A Venturi instrument can be used to measure the volumetric flow rate, Q, using the Bernoulli principle. Since

$$Q = v_1 A_1 = v_2 A_2$$

$$p_1 - p_2 \cdot h1 = \frac{\rho}{2} \left(v_2^2 - v_1^2 \right) \tag{4}$$

then

$$Q = A_1 \sqrt{\frac{2}{\rho} \cdot \frac{(p_1 - p_2)}{\left(\dfrac{A_1}{A_2}\right)^2 - 1}} = A_2 \sqrt{\frac{2}{\rho} \cdot \frac{(p_1 - p_2)}{1 - \left(\dfrac{A_2}{A_1}\right)^2}}$$

[5]

where:

v = velocity in m/sec. v_1 is the slower velocity, where the opening is wider, and v_2 is the fastest, where the opening is narrower

ρ = density of air in kg/m^3

A = area of the wider (A_1) and narrower (A_2) openings

h = height difference between A_1 and A_2.

Cool tower ventilation is only effective where outdoor humidity is very low. The following expression for the air flow induced by the column of cold air pressurising an air supply is based on a form developed by Thompson (1995)[2], with the coefficient from data measured at Zion National Park Visitor Center (see Figure 6.11). This tower is 7.4m tall, 2.4m square cross-section and has a 3.1m^2 opening.

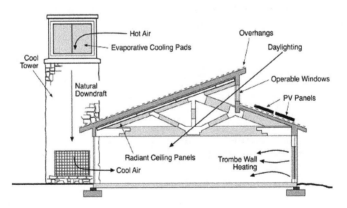

Figure 6.11 Cross-section of Zion National Park Visitor Center, Utah (latitude 37.3°N), showing the use of a cool tower integrated into the building

Source: Creative Commons public domian from the US Department of Energy, P. Torcellini, R. Judkoff and S. Hayter, National Renewable Energy Laboratory. http://www.nrel.gov/docs/fy02osti/32157.pdf

$$Q_{cool} = 0.49 \cdot A \cdot \left[2gh \, \frac{\left(T_{db} - T_{wb}\right)}{T_{db}} \right]^{1/2}$$

[6]

where:

Q_{cool} = volume of ventilation rate (m^3/s)

0.49 = empirical coefficient calculated with data from Zion Visitor Center, UT, which includes humidity density correction, friction effects and evaporative pad effectiveness

A = free area of inlet opening (m^2), which equals area of outlet opening

g = 9.8 (m/s^2), the acceleration due to gravity

h = vertical distance between inlet and outlet midpoints (m)

T_{db} = dry bulb temperature of outdoor air (K), note that 27°C = 300 K

T_{wb} = wet bulb temperature of outdoor air (K).

The air flow from each source can be combined in a root-square fashion as discussed in the ASHRAE Handbook, Fundamentals. The presence of mechanical devices that use room air for combustion, leaky duct systems or other external influences can significantly affect the performance of natural ventilation systems.

6.10   Pre-cooling air with ground-source intakes

Also known as an earth-air tunnel, this is a traditional feature of Islamic and Persian architecture. It utilises pipes buried a few metres down, or underground tunnels, to cool (in summer) and to heat (in winter) the air passing through them. They can lower or raise the outside replacement air temperature for rooms which are buffer zones between the interior and exterior temperatures. Chief points:

• the air may be drawn in by natural convection created by turbine vents, with no fans, thermostat or electricity, from a series of intakes outside the building sited to face the prevailing wind;

• the pipes must be in contact with the soil. They could, alternatively, pass through a buried water tank;

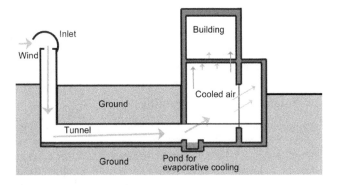

Figure 6.12 Earth-air tunnel principle

Source: Author

- the cooled air enters the building and is drawn up and through it using the stack effect and controlled window openings on different floors;
- in larger buildings this may require a whole-building energy management system to control the louvre opening;
- factors to be taken into account include: the surface area of pipe, length, depth below ground, dampness of the earth, humidity of inlet air velocity;

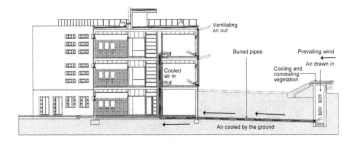

Figure 6.13 Sectional view of the Solar XXI building in Lisbon, Portugal (latitude 38.7°N), showing distribution of the buried air pre-cooling system using an earth-air tunnel network (there is a row of these along the building)

Source: International Energy Agency (IEA)

- in humid climes water will condense inside the pipes and must be allowed to drain away.

In general, with passive cooling, if free ventilation is not feasible or adequate on its own, then mechanical ventilation is used as a supplement, but will be more efficient if it takes place only at night.

6.11   Solar chimneys for cooling

Solar chimneys are employed where the wind cannot be relied upon to power a wind tower. The chimney's outer surface (painted black and glazed) acts as a solar collector, to heat the air within it. (It must therefore be isolated by a layer of insulation from occupied spaces.) This heated air rises, sucking out hot air from inside the building. Its power depends on:

- the size of the collector: the larger it is, the more heat is collected;
- the size of the inlet and outlet: an inverted 'funnel' is best, with the narrow opening at the top;
- the vertical distance between the inlet and outlet: a longer distance creates greater pressure differentials.

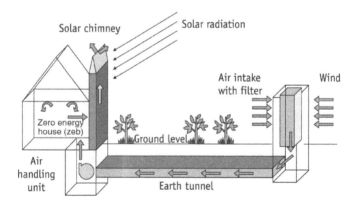

Figure 6.14 Schematic diagram of a solar chimney working together with a wind tower and earth tunnel system suitable for a hot location

Source: Kishore, 2009[3]

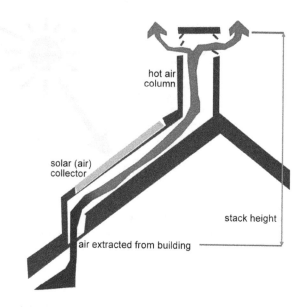

Figure 6.15 The solar chimney principle

Source: Author

Figure 6.16 The solar chimney principle applied on a glass-walled stair well at the corners of a tax office building in Nottingham, England (latitude 53°N). The roof-tops open when required to release hot air

Source: Author

Table 6.3 The effect of vault orientation on seasonal direct solar radiation.[4] CSR = cross-section ratio. This is the ratio between vertical height of the vault and the horizontal width

Orientation	Season	Loss of direct solar radiation (%)				
		CSR1 = 0.5	CSR1 = 0.8	CSR1 = 1	CSR1 = 1.25	CSR1 = 2
W–E	Summer	12.4	20.1	23.9	29	37.8
	Winter	9.8	17	19.6	23.2	30.4
N–S	Summer	17	28.6	35.1	42.1	56.4
	Winter	6.3	7.1	8	8.9	10.7
NE–SW	Summer	14.7	23.9	29.3	34.8	45.6
	Winter	8.9	13.4	16	18.8	24.1
NW–SE	Summer	14.7	23.9	29.3	34.8	45.6
	Winter	8.9	13.4	16	18.8	24.1

6.12   Vaulted roofs and domes

Vaulted roofs and domes dissipate more heat by natural convection than flat roofs. They give greater thermal stability and lower daytime temperature. The best orientation requires that the vault form receives maximum daily solar radiation in winter and minimum in summer. A north–south axis orientation for a vaulted roof is better for winter heating, receiving the minimum direct solar radiation in the summer, while an east–west axis orientation will maximise summer heating, receiving the most irradiation in the morning and evening. The results are summarised by example for a 30° latitude site in Table 6.3.

6.13   Trombe wall effect for cooling and heating

A double-skin façade is employed, the outer skin of which can be glass or PV panels. The cavity between the array and the wall possesses openings to indoors and outdoors at both high and low levels. Heat from the rear of the outer skin causes a convective flow (PV panels will give off more heat and need cooling to operate more efficiently anyway). In winter, either the lower outer or lower inner and the upper inner openings are open. This takes air either from outside or from inside on the lower floor, to be heated and drawn into the building at the top, where it cools and circulates down. In summer, the

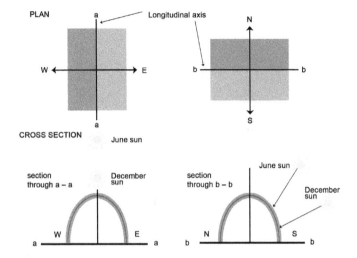

Figure 6.17 The effect of vault orientation on received seasonal direct solar radiation

lower inner and the upper outer openings are opened to let the interior warm air outdoors.

6.14   Evaporative cooling

Evaporative cooling is used in times of low or medium humidity. As water is evaporated (undergoing a phase change to water vapour), heat is absorbed from the air, reducing its temperature. When it condenses (another phase change), energy is released, warming the air. This is the same for all phase change materials. Two temperatures are important when dealing with evaporative cooling systems, used to calculate the relative humidity:

- *dry bulb temperature*: the commonly understood air temperature, measured by a regular thermometer exposed to the air stream;
- *wet bulb temperature*: the lowest temperature that can be reached by the evaporation of water only and a measure of the potential for evaporative cooling.

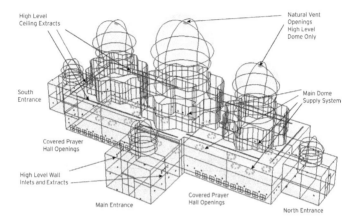

High Level
Ceiling Extracts

Natural Vent
Openings
High Level
Dome Only

South
Entrance

Main Dome
Supply System

Covered Prayer
Hall Openings

High Level Wall
Inlets and Extracts

Covered Prayer
Hall Openings

Main Entrance

North Entrance

Figure 6.18 A design sketch of the Sheikh Zayed Grand Mosque in Abu Dhabi (latitude 24.5°N). Prior to construction, thermal analysis was conducted to determine the required level of thermal comfort using a building information model (BIM). An initial model determined the impact of a conventional air-conditioning system. This was then refined with fresh air supply from the central air-conditioning induction system using the stack effect and cross-ventilation with extract and supply louvres located at high level around the central support columns. A computational fluid dynamics (CFD) analysis of the main prayer hall, which can accommodate 9,000 people during Ramadan, modelled thermal comfort and air flow patterns

'Humidity' describes how much water is already in the air relative to the amount it is capable of holding. With evaporative cooling, the amount of sensible heat absorbed depends on the amount of water that can be evaporated.

Evaporative cooling can be direct or indirect; passive or hybrid.

- *Direct*: the humidity of the cooled air increases because air is in contact with the evaporated water – can be applied only in places where relative humidity is very low.
- *Indirect*: evaporation occurs inside a heat exchanger and the humidity of the cooled air remains unchanged – used where humidity is already high.

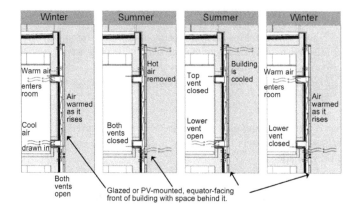

Figure 6.19 Sectional view of the Solar XXI building in Lisbon, Portugal, showing the method of operation of the heat output of wall-mounted PV modules to supplement ventilation (it could be glazing). The lower opening can either be open to the room to provide ventilation or to outside to provide cooling for the PV panels only

Source: International Energy Agency (IEA)

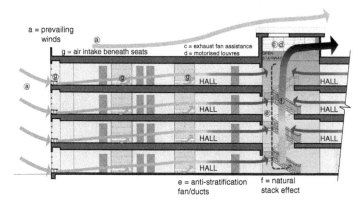

Figure 6.20 Natural ventilation in a large building with an atrium that doubles as a stairwell

Source: Adapted from an illustration for a LEED Gold standard project, UC Davis Tercero 2 Student Housing, by Mogavero Architects (2013), original image licensed under a Creative Commons Attribution Licence. See http://bit.ly/2t0r1HF

- *Passive*: where evaporation occurs naturally; incoming air is allowed to pass over surfaces of still or flowing water, such as basins or fountains.
- *Hybrid*: where mechanical means are deployed to control evaporation.

Passive direct cooling also can utilise water falling vertically along guides where air enters the building, and as a design element in the building façade.

6.14.1 *Passive downdraft evaporative cooling (PDEC)*

A downdraft tower contains cellulose pads at the top onto which water is distributed at intervals or dripped, using micronisers or nozzles. The water is collected at the bottom into a sump and recirculated by a pump. Alternatively, the pressure in the supply water line may be used. Inside, a column of cool air falls at a rate, depending on the efficiency of the device, tower height and cross-section, as well as the resistance to air flow. At the Torrent Research Centre in Ahmedabad inside temperatures of 29–30°C were recorded when outside temperatures were 43–44°C, with six to nine air changes per hour on different floors, by using this method.

6.14.2 *Roof surface evaporative cooling (RSEC)*

In a dry tropical climate, roof surfaces are sometimes cooled by spraying water over suitable water-retentive materials spread on the roof. The wetted roof temperature may be 15°C less than ambient air of 55°C, but the system is only appropriate in areas where water is not scarce. Variations include:

- *roof pond*: a shallow water layer over a highly conductive flat roof. The pond is covered in daytime with a highly reflective, well-insulated layer that is automatically withdrawn at night, when it potentially lowers room temperature below up to 20°C. By keeping the pond open during night, the water is cooled by nocturnal cooling. Can be used for heating during the winter by operating in reverse.

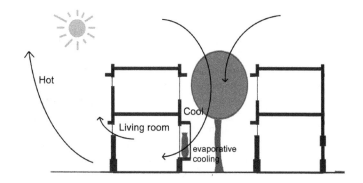

Figure 6.21 Using trees and a courtyard to provide cooling, together with cross-ventilation and evaporative cooling from a water source for the air entering

Source: Author

- *diode roof*: a corrugated sheet-metal roof holding polyethylene bags coated with white titanium oxide containing pebbles and water. Daytime-evaporated water condenses inside at night, permitting 4°C of cooling below the minimum air temperature.

6.15   Phase change materials (PCMs)

A phase change material (PCM) is one which melts and solidifies at a certain temperature, such as wax or salt. This makes it capable of storing and releasing large amounts of energy. Latent heat is absorbed when the material changes from solid to liquid and released when it solidifies.

In buildings, they are useful in places with a day–night temperature variation. In the daytime, incoming hot external air is cooled by the PCM storage module, which absorbs and stores its heat by melting. At night-time the substance reverts to solid form, releasing its heat by being cooled by the now cooler external air. Commercially available PCMs are chosen based on the temperature of their phase change relative to that required in the space to be moderated. They come in four categories:

1. *eutectics*: solutions of salt in water with a phase change temperature below 0°C (+32°F);
2. *salt hydrates*: specific salts able to incorporate water, which crystallise during the freezing process, normally above 0°C (+32°F);
3. *organic materials*: which tend to be polymers composed primarily of carbon and hydrogen. These mostly change phase above 0°C (+32°F) and can be as simple as coconut oil, waxes or fatty acids;
4. *solid–solid materials*: there is no visible change in the appearance of the PCM (other than a slight expansion/contraction) during phase change and therefore no problems associated with handling liquids, e.g. containment, potential leakage, etc.

For example, an air-conditioning device may be integrated with a heat exchanger and use a PCM based on expanded natural graphite, for heating or cooling a space. The melting point of this particular PCM package is 20°C and its heat capacity is 30Wh/kg. Its rated air flow is around 160m³/h. For cooling loads of up to 50W per square metre of floor area, a rough figure for the mass of the material required is 5.5kg per square metre. Thermal modelling will help size the system correctly.

Plasterboard is available that is integrated with small plastic capsules with a core of wax – a PCM – inserted during fabrication. The melting temperature of the wax can be defined during manufacturing. If the room temperature rises above this melting point (around 21–26°C), the wax melts, absorbing as it does so the surplus room heat. Conversely, when the room temperature falls, the wax sets, yielding heat to the internal atmosphere. Office spaces in India have installed this plasterboard in the ceiling.

PCMs may be used in combination with night ventilation to ensure the wax solidifies at night, otherwise it will not work during the day. The wax capsules are delivered from the manufacturer as liquid dispersion or powder.

6.16   Night cooling

Night ventilation, or night flushing, relies upon keeping windows and other openings closed during the day but open at night to flush warm air out of the building and cool thermal mass which has heated up during the day.

It requires significant temperature differences between day and night-time (which must be below 22°C/71°F) and some wind movement. It is necessary to open up pathways for wind ventilation and stack ventilation through the night. At the beginning of the day the building is closed and sealed to prevent warm air from outside entering. It is necessary for large areas of thermal mass inside not to be covered. Opening and closing may have to be on a timer. Security precautions should be taken for ground-level openings.

6.17 Earth covering of buildings

At depths beyond 4–5 metres, soil temperature remains almost constant throughout the year. Thus, partially earth-covered buildings will obtain both cooling (in summer) and heating (in winter).

6.18 Desiccant cooling

Desiccant cooling is effective in warm and humid climates. A desiccant is a hygroscopic subtance that absorbs moisture from the air, helping to cool it. Desiccants are silica gel, alumina gel and activated alumina, or liquids like triethylene glycol. They may be supplemented by mechanical dehumidifiers driven by solar thermal power. Desiccant cooling ventilation systems do not need ozone-depleting refrigerants and car treat large humid volumes of air. Air from the outside enters the unit to be dried adiabatically before entering the living space. The desiccants are regenerated by solar energy. Desiccant cooling may be used in conjunction with evaporative cooling to adjust air temperature to the required level.

6.19 Solar-powered absorption chillers

These chillers can dehumidify and cool air using solar collectors (evacuated tube or parabolic) as a source of energy. An active rather than a passive solar technology, they are described in the companion book to this volume, *Solar Energy Pocket Reference*. Their main advantages are:

- use of renewable energy;
- low electricity consumption;
- compact and space-saving with outdoor installation;

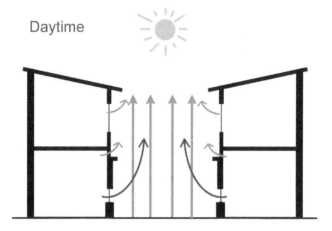

Daytime

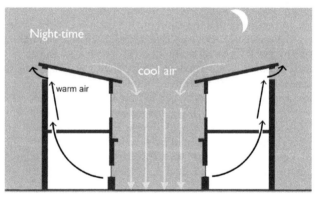

Night-time

cool air

warm air

Figures 6.22 The use of a courtyard with roofs sloping into the courtyard, and night-time ventilation, to provide day and night-time cooling. In the daytime, hot air rising in a courtyard draws cooler air from inside the building. During the night, cool air falling from the roof into the courtyard is drawn into the building to replace the warm air rising out of openings at the top on the outside. Note: the roofs must be white or reflective, insulated beneath and slope inwards to the courtyard

Source: Author

- simple to install, requiring only connection of pipes for water, conditioning system and makeup water to compensate for losses by evaporation from the cooling tower;
- modulation according to the thermal load: frigorific power can be modulated from 20 to 100 per cent;
- high efficiency (coefficient of performance = c.0.7);
- long life-cycle and low maintenance due to minimal moving mechanical components and compressors;
- no noise or vibration;
- low fixed costs.

6.20   Green roofs and walls

Green roofs and external walls help to create a local micro-climate which can not only cool the surrounding urban air, but also provide shade and improve air quality. They will contribute to an increase in biodiversity and help with natural regeneration. Vegetation must be selected with regard to local climate and species, ease of maintenance (watering and leaf shedding, weeding, etc.), and the weight load of both the growing medium and grown plants.

Notes

1 From a National Renewable Energy Laboratory article, 'Natural Ventilation', by Andy Walker, 15 June 2010, available at: http://www.wbdg.org/resources/naturalventilation.php
2 Thompson, L., *NatVent Project*. Available at: http://projects.bre.co.uk/natvent/
3 Kishore, V.V.N., *Renewable Energy, Engineering and Technology: A Knowledge Compendium*, Tata Energy Research Institute, New Delhi, 2009.
4 Mashina, G. A. and Gadi, M. B., 'Calculating direct solar radiation on vaulted roofs using a new computer technique', Nottingham University Conference Proceedings, 2010. Available at: http://www.engineering.nottingham.ac.uk/icccbe/proceedings/pdf/pf196.pdf

7   Passivhaus

The Passivhaus standard (see www.passiv.de) for new and existing buildings is the most reliable method for achieving low and zero carbon buildings. Passive house (German: Passivhaus) is a rigorous, voluntary standard for energy efficiency in a building. It dates back to 1988 and research by Bo Adamson of Lund University, Sweden, and Wolfgang Feist of the Institut für Wohnen und Umwelt (Institute for Housing and the Environment), Germany.

Passivhaus is a way of ensuring that the twin aims of passive solar building – to reduce fossil fuel energy use and increase comfort – are met empirically. The key distinguishing factor of Passivhaus is the rigour applied in the design process to attaining and measuring a specific maximum level of energy use. This takes its inspiration from the field of energy management, which operates under the mantra: 'what gets measured gets saved'. Only by quantifying energy use can we be sure that hoped for energy savings are actually achieved in practice. Passivhaus tries to anticipate this post-occupancy use of the building by taking highly targeted measures in the design, construction and certification stages to give occupants a strong chance of minimising their energy use.

Therefore, Passivhaus buildings in general need no conventional heating or cooling system, making them very cheap to run. Some type of heating/cooling is usually required and this is usually distributed through a low-volume heat recovery ventilation system that also maintains air quality.

The first Passivhaus row of terraced houses was built in Darmstadt, Germany, in 1990 and occupied by the clients the following year. In September 1996 the Passivhaus Institut was founded in Darmstadt to promote and control the standards. Since then, tens of thousands of

Passivhaus structures have been built. In addition, many existing buildings have been renovated to Passivhaus standard.

Passivhaus Institutes have been founded in many countries, including the USA. The principle continues to spread and be adapted to the differing cultural and climatic conditions of other countries. As a result the techniques to achieve Passivhaus standard have expanded considerably and vary similarly. Passive houses are now found as far apart in latitude as Qatar and Scandinavia. It is often possible to construct them for the same cost as for normal building standards in most countries.

7.1   The Passivhaus standard

The standard stipulates a way of analysing all of the heat gains and losses within a building, and modelling improvements through the use of software to achieve the optimum result cost efficiently. The standard requires that:

1.   the space heating or cooling requirement must not exceed $15kWh/m^2/y$;
2.   total primary energy use (of all appliances, lighting, ventilation, pumps, hot water) must not exceed $120kWh/m^2/yr$ (38039 $Btu/ft^2/y$);
3.   building fabric U-values must be less than 0.15 W/m^2K;
4.   the air change rate must be less than or equal to 0.6 air changes per hour, under test conditions. This is 0.6 times the building volume per hour at 50 Pa (N/m^2) and < 1.0 ACH-1 @ 50 Pa for EnerPHit[1] refurbishment projects;
5.   alternatively, when looking at the surface area of the enclosure, the leakage rate must be less than 1.42 litres or 0.05 cubic feet per minute;
6.   thermal comfort must be achieved for all living areas year-round, with not more than 10 per cent of the hours in any given year over 25°C;
7.   there should be no thermal bridges, meaning that any linear (two-dimensional) thermal bridges should have a psi[2] (ψ) value of ≤ 0.01 W/mK.

The (lower) standard for a refurbished building is the EnerPHit Standard, where the space heating energy requirement is

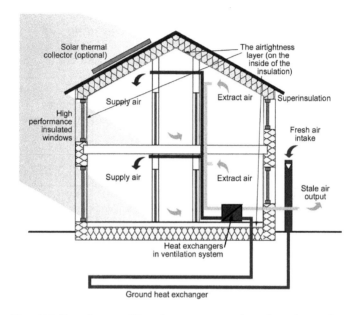

Figure 7.1 Some features of Passivhaus: continuous thermal envelope and airtightness layer; bringing in fresh air from outside via cooling or heating from a ground heat exchanger; mechanical ventilation with heat recovery

Source: Adapted from a Creative Commons-attributed image produced by the Passivhaus Institute

$25\text{kWh/m}^2/\text{y}$, with an airtightness of one building volume air change an hour. This is often applied to Passivhaus retrofits/refurbishments.

Design is usually assisted by use of the Passivhaus/Passive House Planning Package (PHPP), a computer-based tool, which is purchased from either the German Passivhaus Institute (see below) or your national institute, if it exists. Climate information for different locations is available as plug-ins or alternative versions to support this. PHPP is rapidly gaining credence internationally because of the simplicity and flexibility of its approach compared, say, to national building regulations.

Table 7.1 Summary of requirements of Passivhaus standard

Building energy performance	
Specific heating demand	≤ 15kWh/m².yr
or Specific peak load	≤ 10 W/m²
Specific cooling demand	≤ 15kWh/m².yr
Primary energy demand	≤ 120kWh/m².yr
Elemental performance requirements	
Airtightness	≤0.6 ac/h (n50)
Window U-value	≤ 0.80 W/m²K
Window installed U-value	≤ 0.85 W/m²K
Services performance	
MVHR heat recovery efficiency	≥ 75 per cent*
MVHR electrical efficiency	≤ 0.45 Wh/m³
Thermal and acoustic comfort criteria	
Overheating frequency > 25°C	≤ 10 per cent of year
Maximum sound from MVHR unit	35 dB(A)
Maximum transfer sound in occupied rooms	25 dB(A)

Note: * MVHR efficiency must be calculated according to Passivhaus standards not manufacturer's rating

7.2 Strategies

The strategies available to achieve the standard address thermal bridging, airtightness and all energy use. The requirements imply the following features to achieve them:

- passive preheating or cooling of fresh air: brought in through underground ducts that exchange heat or coolness with the ground;
- MVHR (mechanical ventilation with heat recovery) – removing heat from the expelled air and, if required, transferring it to the incoming air. It is possible to transfer over 80 per cent of the heat in the ventilated exhaust air to the incoming fresh air;
- hot water supply using renewable energy: solar collectors, biomass, CHP or heat pumps powered by renewable electricity;
- energy-saving appliances: ultra-low energy LED lighting, refrigerators, technology and so on.

7.3　Windows and doors

With Passivhaus, doors and windows are independently certified. The Passivhaus standard is that a glazing unit (of standard size: 1.24 × 1.48m) has a whole window U-value of ≤ 0.80 W/m²K and can achieve a U-value ≤ 0.85 W/m²K (below 4755 Btu/ft²/yr) when installed. For doors, the installed U-value should be ≤ 0.80 W/m²K. Airtightness is assured with multiple continuous airtight seals, used in conjunction with a robust gearing system. The standard in the UK is ≤ 2.25 m³/hm at Q(100 Pa). Solar transmittance values for the glazing will vary according to the latitude. In temperate and northern zones it may be g-value ≥ 0.5.

Passivhaus windows combine triple-pane insulated glazing with a good solar heat-gain coefficient, low-emissivity coatings, sealed argon or krypton gas filled inter-pane voids, 'warm edge' insulating glass spacers and air-seals. When unobstructed and equator-facing, the solar heat gains are, on average, greater than the heat losses, even in mid-winter.

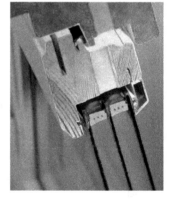

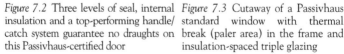

Figure 7.2 Three levels of seal, internal insulation and a top-performing handle/catch system guarantee no draughts on this Passivhaus-certified door

Figure 7.3 Cutaway of a Passivhaus standard window with thermal break (paler area) in the frame and insulation-spaced triple glazing

7.4   Airtightness

Besides this and superinsulation, the main key to the success of Passivhaus design and build is a very high level of airtightness compared to conventional construction.

Airtightness enables ventilation to be managed proactively. Most air exchange with the outside happens through controlled ventilation with heat recovery through a heat exchanger to minimise heat loss or gain, depending on climate. This degree of airtightness

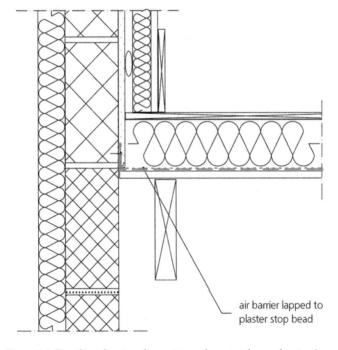

air barrier lapped to
plaster stop bead

Figure 7.4 Detailing showing the position of an airtightness barrier (grey dotted line) beneath the insulation under a suspended floor, lapping up the side to meet plaster/plasterboard. On the inside of the plaster is more insulation. Insulation is also on the outside of the building. Together this makes the building thermally insulated and airtight. Diagrams like this are free from the enhanced construction details (ECDs) section of the Energy Saving Trust website. See http://bit.ly/2fKMDp3

Source: EST

requires a continuous air barrier system around the building envelope. This barrier is a combination of interconnected materials, assemblies, flexible sealed joints and components of the building enclosure or skin, and separations between conditioned and unconditioned spaces. It is connected to the foundation walls and basement slabs. Each seal must be made securely, to last for many years. Impeccable scrutiny must be given to: continuity, structural support, air imperme-ability and durability.

Windows are therefore frequently only openable under certain conditions by occupants, which does not suit everyone. Careful management of moisture and dew points is needed and so knowledge of how humidity, temperature and pressure behave within the build-ing's walls is important. Success in avoiding damp or condensation is achieved through air barriers, careful sealing of every construction joint in the building envelope and sealing of all service penetrations. See section 7.7 below for more information. Attention to detail is vital during construction, particularly in ensuring that all seals will last as long as possible.

The airtightness layer typically uses specialist membranes, boards and tapes. Exactly which depends upon the construction type. The airtightness barrier may consist of:

- *vapour permeable (and hygroscopic) materials*: vapour perme-able membranes, intelligent membranes, lime, concrete, timber, hempcrete, bricks, stone, plaster, mineral (rock and slag) wool;
- *non-permeable materials*: PVA and vinyl paint, when used as a coating on plaster, metal, glass, PVC, plastic, plastic foams, XPS, EPS, phenolic foam insulants.

Other pressure differentials within buildings should be controlled by:

- compartmentalising and sealing garages under buildings with airtight walls and vestibules at building access points;
- compartmentalising spaces under negative pressure, such as boiler rooms, and providing makeup air for combustion;
- sealing the interior and removing thermal bridging to avoid convection currents within enclosed assemblies caused by a cold side contacting air on the warm side of insulation, in order to combat mould formation in insulated basements.

Figure 7.5 Cross-sections of inside and outside of a pitched roof and wall showing (in darkest shades) the use of tapes to ensure airtightness around thermal envelope penetrations such as flues and windows. The illustrations also show the use of the vapour and moisture layers and insulation to prevent thermal bridging

Airtightness is defined by an n50 test measurement ('blower door test') which combines both under and over pressurisation tests. The air leakage at a pressure of 50 pascals must be no greater than 0.6 air changes per hour (0.6 ACH @50 Pa) based on the building's volume, or 0.05 CFM50/sf (cubic feet per minute at 50 Pa, per square foot of building enclosure surface area). It is recommended that there are two tests: one while the airtight barrier is still exposed and the final test upon completion.

Siting the airtight barrier within the construction build reduces the need for repeated penetrations of the building envelope for services such as cabling, pipes, etc. Where these penetrations do occur, proprietary airtight gaskets and grommets are available.

Building contractors must be made aware of the imperative to maintain the airtight barrier all around the building. For large buildings, separate airtightness zones are often specified.

7.4.1   Detailing

Well-considered Passivhaus details with continuous insulation and airtightness combined with minimal or no thermal bridging must be supplied by the architect. There will be drawings for all junctions: with windows, ground level, midfloor and roof level. The airtight layer location, the thermal envelope and the materials should be easy to identify from the drawings.

The best components can become ineffective if not installed correctly. Installers must be educated and supervised so they do not apply their possibly habitual practices of, for example, spraying foam and injecting silicone sealer around the expensive triple-glazed windows. This would at some point shrink and create a bypass for air to flow from or to the outside.

Construction process instructions are helpful. Here is an actual example, for a floor-wall junction:

> Connect the vapour barrier with an air-tight seal. Perform the blower door test before assembling the floor structure to seal existing leaks. Be careful to avoid ruptures and other leaks in the sealing layers in the valleys between the floor slab and the wall since post-construction repairs are difficult and complex. The gypsum plasterboard panels on the inside ought not touch the

Figure 7.6 Installation of tongue-and-groove wax-impregnated wood fibre-board cladding over wood fibre batts on timber frame. The cladding was afterwards rendered with lime plaster to make the walls breathable

floor slab due to the possibility of condensation damage. A lower thermal insulation layer is specified.[3]

7.5 Passive House Planning Package (PHPP)

The Passive House Planning Package (PHPP) is the design tool produced by the Passivhaus Institute to model the performance of a proposed Passivhaus building. The Passivhaus must be modelled and verified using the PHPP while using the appropriate regional climatic dataset. The PHPP is intended for use by anyone involved in the design of a Passivhaus.

The PHPP consists of a user manual and CD containing the PHPP 'Excel-based' design tool and is compatible with both PC and Mac computers. To use the PHPP software, information needs to be entered about the planned building and the location. Parameters such as window angles, sizes and shading can be changed to see what effect this has on the energy demand. This can save much time and energy and avoid costly mistakes later. PHPP contains a series of tools for:

Figure 7.7 Sample details for mini-mising thermal bridging on differ-ent types of roof: cold roof, flat roof with cavity wall and flat roof with timber wall

- calculating energy balances;
- calculating U-values;
- designing comfortable ventilation;
- calculating the heating and cooling load;
- summer comfort calculations;
- and many other useful tools for reliable design of Passivhaus dwellings.

The heat loss areas and thermal bridges are calculated relative to the external boundary layer and, with attention to detailing, negative psi values may even be obtained. This means that insulation is so good that the two-dimensional heat flow through a junction is less than the respective one dimensional heat flows. All significant ther-mal bridges must be recorded, requiring accurate modelling using

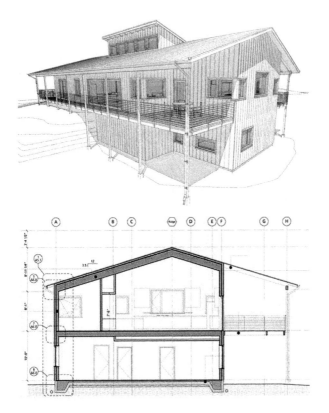

Figure 7.8 A timber frame Passivhaus home design and a cross-section showing the location of the airtightness layer (dark line). The assembly was specified as follows:

Roof assembly:

- roofing material
- moisture barrier
- wood fibreboard
- solid insulation w/ taped joints
- continuous moisture barrier
- plywood decking, structural
- timber roof framing, structural
- w/blown-in cellulose insulation

- plasterboard.

External wall assembly:

- rainscreen
- two layers of rigid insulation
- all joints staggered and taped
- moisture barrier
- all penetrations taped and sealed
- structural plywood sheeting
- advanced timber framing

- w/blown-in cellulose insulation
- plasterboard.

Slab floor construction:

- flooring finish
- concrete w/underfloor heating
- vapour and airtight layer
- solid insulation
- sand/gravel
- compacted earth.

specialist software. Therefore, the first step is to identify them all and design them out. Often the problematic areas are at wall and roof junctions and around the windows. The following data are entered:

1. climate data, number of occupants, building type;
2. degree day data;
3. enclosed volume;
4. wall materials composition and thickness to calculate U-values;
5. external dimensions of building;
6. heat losses via ground;
7. window type and windows, orientation;
8. shading;
9. ventilation, including air flow rate;
10. energy balance;
11. thermal bridges;
12. summer ventilation and shading;
13. annual heating/cooling demand and heating/cooling load;
14. hot water, distribution, renewable energy, primary energy, boiler.

In PHPP, to find the heating (or cooling) degree hours, G_t, the temperature difference between inside and outside is calculated on an hourly basis. The unit of measurement is kKh/a (kilo-Kelvin hours per year). The specific space heat demand, Q_H, is arrived at through the following steps:

1. calculate heat losses: transmission losses Q_T plus ventilation losses Q_V;
2. calculate heat gains: solar gains Q_s plus internal heat gains Q_I;
3. subtract gains from losses; the difference is the specific space heat demand:

$$Q_H = Q_T + Q_V - \eta^* (Q_S + Q_I) \qquad [1]$$

where η is the utilisation factor for heat gains – for example, 84 per cent.

The *transmission heat losses* are calculated from the area of the thermal envelope (A) multiplied by the U-value (U) multiplied by the temperature-correction factor (T), multiplied by the heating degree hours (G), plus a factor for the thermal bridges ψ:

$$Q_T = \Sigma\left(U_I + A_i + G_{T_i}\right) + \Sigma\left(\Psi_i + I_i + G_{T_i}\right) \qquad [2]$$

The *ventilation heat losses* are calculated by multiplying the ventilated volume by the equivalent η change, by the heat capacity and air, by the heating degree hours:

$$Q_v = V^* n_{equiv.}^* C_p \, \rho^* G_t \qquad [3]$$

The *solar gains* are a reduction factor multiplied by the g factor of the windows, the window area and the global irradiation figure (in kWh/$(m^2 a)$):

$$Q_S = r^* g^* A^* G \qquad [4]$$

The *internal heat gains* are the extent of the heating period (in days per year) multiplied by the specified internal heat gains (in W/m^2) multiplied by the treated floor area (in square metres):

$$Q_I = t_{Heat}^* q_i^* A_{TFA} \qquad [5]$$

The *specific heat/cooling demand* is the transmission losses and ventilation losses minus η (the utilisation factor for heat gains – for example, 84 per cent) times the solar gains plus internal heat gains:

$$Q_H = Q_T + Q_V - \eta^*(Q_S + Q_I) \qquad [6]$$

To calculate the heating or cooling load in a passive house:

1. calculate heat losses in worst possible weather;
2. calculate heat gains in worst possible weather;
3. the difference is the heating load in the worst possible weather.

$$P_H = P_T + P_V - (P_S + P_I) \qquad [7]$$

The reverse is done for cooling loads in the hottest conditions.

Sources of climate data are found in Chapter 10. Note: local micro-climates may still perform differently. In addition, occupant

Height of the Shading Object	Horizontal Distance	Window Reveal Depth	Distance from Glazing Edge to Reveal	Overhang Depth	Distance from Upper Glazing Edge to Overhang	Additional Shading Reduction Factor
m	m	m	m	m	m	%
h_{Hori}	d_{hori}	o_{Reveal}	d_{Reveal}	o_{over}	d_{over}	r_{other}
		0.23	0.105	0.23	0.105	
		0.18	0.105	0.23	0.105	
6.71	8.00	0.23	0.105	0.23	0.105	
		0.23	0.105	0.23	0.105	
		0.23	0.105	0.23	0.105	
		0.23	0.105	0.23	0.105	
		0.23	0.105	0.52	0.85	
3.75	8.00	0.23	0.105	0.23	0.105	
		0.23	0.105	0.52	0.85	
3.79	8.60	0.23	0.105	0.23	0.105	

Figure 7.9 Example of a PHPP page. This one is for shading calculations

behaviour can influence building performance much more than the local climate. Identical passive buildings in the same location can vary in performance by 50 per cent due to occupant behaviour. The provision of manuals, 'how-to' videos, apps, etc. for reference by successive occupants is recommended.

7.6   Recommendations

The Passivhaus Trust in the UK issues a booklet of recommendations for designers and installers.[4] The following is a short summary.

Design stage:

- Passivhaus is an integrated part of the design approach so must be understood by all members of the design team from the early stage;
- success depends on high levels of fabric performance and carefully controlled solar gain. It does not depend upon either light or heavy thermal mass;
- do minimise complexity to reduce the chance of error;
- alternative choices for the design are explored using the PHPP software to test design decisions. This will include exploring the impact of the building's form-factor ratio;
- training is recommended in the use of this software. If in doubt, employ a certifier to help enter the data;
- use a robust quality assurance process. Certification usually starts pre-construction;

- ensure the correct weather data set for the location is used and entered;
- remember that the heat loss areas are measured to the outside of the thermal envelope;
- ensure repeating thermal bridges are entered correctly into the software;
- be critical of manufacturers' data concerning performance of their products;
- window data also needs to be entered accurately. Use PHPP to assess the whole window component performance, not just the window's U-values (R-values);
- the ventilation rate will vary according to the norm for the country where the building is located;
- on overheating, the upper limit is to keep the building below 25°C for <10 per cent of the year, and best practice is <5 per cent of the year;
- be aware that fresh air and exhaust ducts for MVHR are outdoor spaces within the thermal envelope so can add significantly to heat loss. The use of multiple MVHRs requires the use of the Additional Vent sheet;
- don't be over-reliant on large amounts of glazing to achieve success as this makes success dependent upon the weather – which is not reliable;
- entering shading data correctly requires a lot of care. Specialist software is available to do this such as IES, TAS or Ecotect;
- consider which construction system is best suited to achieve Passivhaus requirements. Avoid mixing different systems within one building;
- choose an appropriate contractor who understands Passivhaus;
- details and junctions need to be designed bearing in mind construction and assembly on-site under realistic conditions to ensure they will end up airtight;
- with steel and concrete frames, address potential thermal bridges everywhere and especially with the connection of the columns to foundations. Locate the airtightness line within the wall where possible;
- timber frame is generally easier to build on a concrete raft floating on EPS insulation;
- with traditional masonry achieving Passivhaus requires very wide insulated cavities between inner and outer walls;

- avoid thermal bypasses caused by air movement around and through the building fabric by locating a breather membrane layer on the outside;
- generally, the more thermal mass that is within the thermal envelope the easier it is to control the possibility of overheating in hot weather. This can be achieved even with a timber frame construction by using a solid ground floor.

Construction stage:

- don't allow the construction schedule to override quality requirements;
- use modern reinforced purpose-made membranes with specialist airtightness tapes to achieve airtightness;
- special techniques are required with masonry construction with cavities to avoid the use of mortar compromising airtightness and insulation. Extra care is needed;
- insulating layers should meet at junctions between elements without any gaps. Apply robust details with adequate construction tolerance;
- don't allow disruptions to the thermal envelope. When they do occur make the thermal resistance in the insulation as high as possible;
- always overestimate construction tolerances;
- appoint a quality assurance and airtightness champion. Don't assume something is sealed, always check;
- for the airtightness test, leave a margin to allow for the leakage often found around services that are installed after the first test of at least 25 per cent, making the targets for the preliminary test: <0.45 ACH-1 @50Pa for new build projects and <0.75 ACH-1 @50Pa for refurbishment projects;
- provide a manual for occupants to illustrate how to successfully manage the building in line with its design.

7.7   Dealing with humidity

The higher the insulation level, the less heat is available for drying out moisture in the building. Additionally, improved airtightness can lead to increases in the indoor air humidity. Together these factors raise the importance of moisture protection of the building components. Tools are available to tackle these risks in the design stage and are important for any passive house designer. Hygrothermal simulations like WUFI® and Delphin are therefore important as they allow realistic calculation of the transient coupled one- and two-dimensional heat and moisture transport in walls and other multi-layer building components exposed to natural weather. WUFI® is an acronym for Wärme Und Feuchte Instationär which, translated, means heat and moisture transiency. See Chapter 10 for more information.

7.7.1 Interstitial condensation

One of the chief challenges is avoiding interstitial condensation. This can occur in solid and cavity wall structures. It happens when pressure and temperature differences force warm humid air through hygroscopic (water absorbing) materials until they reach a point cold enough for it to condense upon a surface (the dew point). It is the result of the inter-action between a complex set of factors. This includes:

- the amount of steam and evaporation and the occupancy level in a given room;
- relative humidity, temperature and pressure both inside and outside (which varies all the time);
- the structure and composition of a wall, floor or ceiling/flat roof (sloping roofs above ventilated lofts are rarely a problem);
- and the surfaces (e.g. paint type), both inside and outside, of a wall.

Each case will be different. In general, walls, floors and ceilings need to be hygroscopic (able to absorb moisture from the air) in order to let the building 'breathe'. The solution can never be simply to block off the possibility of moisture passing through the surface of a wall, ceiling or floor on the inside.

Relative humidity decreases as the internal temperature rises. To have the best chance of eliminating the problem, the layers of materials within the wall, floor or ceiling all need to be three things:

1. hygrothermal (able to absorb more moisture than can it releases at any one time, without harm, until conditions change so that it can be released);
2. porous;
3. progressively more vapour-resistant moving from the warm (inside) to the cold side.

However, this strategy becomes complex in climates where there is significant seasonal variation. The positioning and type of vapour barrier becomes vital. This is where modelling software is required.

7.7.2 Vapour resistance

The vapour resistance of a material is a measure of its resistance to letting water vapour pass through. It is measured by its μ-value ('mu-value'), also known as its 'water vapour resistance factor'. This is governed by the standard ISO 10456, which specifies methods for the determination of declared and design thermal values for thermally homogeneous building materials and products, together with procedures to convert values obtained under one set of conditions to those valid for another set of conditions. These procedures are valid for design ambient temperatures between -30 °C and +60 °C. Here are some examples of the μ-value of materials.

To convert a μ-value to the vapour resistance (MNs/g) of a given material, multiply by its thickness in metres, then divide by 0.2 (g.m/MN.s). For example, for a material with a μ-value of 60 and thickness of 30cm, its vapour resistance is 60 × 0.3 m ÷ 0.2 g.m/MN.s = 1 MN.s/g.

In the USA the unit of measurement for characterising the water vapour permeance of materials is the 'perm', a measure of the rate of transfer of water vapour through a material (1.0 US perm = 1.0 grain/square-foot·hour·inch of mercury ≈ 57 SI perm = 57 ng/s·m^2·Pa). American building codes classify vapour retarders as having a water vapour permeance of 1 perm or less when tested in accordance with the ASTM E96 desiccant, or dry cup method. Materials are categorised as:

- *impermeable* (≤1 US perm, or ≤57 SI perm) (e.g. asphalt-backed kraft paper, vapour-retarding paint, oil-based paints, vinyl wall coverings, extruded polystyrene, plywood, OSB);
- *semi-permeable* (1–10 US perm, or 57–570 SI perm) (e.g. unfaced expanded polystyrene, fibre-faced isocyanurate, heavy asphalt-impregnated building papers, some latex-based paints);
- *permeable* (>10 US perm, or >570 SI perm) (e.g. unpainted gypsum board and plaster, unfaced fibre-glass insulation, cellulose insulation, unpainted stucco, cement sheathings, spunbonded polyolefin or some polymer-based exterior air barrier films).

Using this information, it is possible to calculate the progressive vapour resistance of the various layers of different materials of different thicknesses within a wall/ceiling/floor.

7.7.3 Intelligent membranes

However, the temperature and humidity levels can affect the permeance of a material. This makes it possible to construct vapour curves for them. These plot perms against mean relative humidity. While many materials have variable vapour permeability, the quality of the permeability curve matters in choosing the right materials.

This has led to the development of 'intelligent' or 'smart' membranes. These are a type of vapour control layer which will change its permeability according to relative humidity, pressure and temperature conditions, so it can vary between being a vapour barrier and a breathable membrane. It is generally taped into place over the insulation, on the warm side.

Moisture issues in walls – such as rot, mould and rust – occur at 80 per cent relative humidity (RH) and above. Therefore it is important to choose a variable vapour retarder, or smart membrane, that opens up quickly and as much as possible when relative humidity exceeds 70 per cent, to facilitate inward drying.

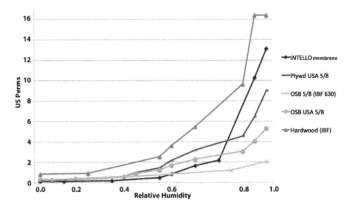

Figure 7.10 An example of vapour curves plotting relative humidity against permeance for different materials, including a proprietary variable vapour permeability membrane, INTELLO

As an example, consider the case of a wall structure that contains a polyethylene vapour barrier on the inside of a board such as OSB (Oriented Strand Board, a timber-product sheet material). If it becomes more humid outside than inside, the vapour drive will be inwards. The vapour is likely to condense on the cold side of the non-breathable plastic barrier and risk causing rot in the board. Replacing the polyethylene membrane with one that is variable in its permeability and that will allow moisture through when relative humidity exceeds 70 per cent can address this. The building can still be air-conditioned to remove the excess moisture.

The same principle can be applied in reverse in the case where it is cold outside and humid inside the building.

Manufacturers of variable vapour permeability membranes now claim that with their materials it is possible to build Passivhaus-certified structures that perform correctly in practically any climate, using exterior vapour retarders such as OSB, taped and sheathed, flat roofs, unvented asphalt roofs and more. Vented roofs and walls in mixed and humid climates are also possible without the use of foam which carry no risk from condensation throughout the year.

7.8   Passivhaus around the world

To promote and further evaluate the Passivhaus standard in various climates, the Passivhaus Institute has issued a number of performance-related studies. 'Passive House in Different Climate Zones' is a study carried out by the Institute to assess the performance of the standard in extreme climates. A graphical representation (Figure 7.12) was used to measure thermal comfort in the projects. Annual hourly operative temperatures and the concurrent relative humidity are plotted against each other. A central shaded area represents the inner comfort zone, covering a wide range of temperatures (20°C to 27°C) and relative humidity levels (30 per cent to 70 per cent). The outer peripheral area represents the extended comfort zone.

Climate data for input into PHPP software are now available for a wide range of locations and many are listed in Chapter 10.

Figure 7.11 Air intakes for ventilation and cooling outside the Energon passive office building in Ulm, Germany

Source: International Energy Agency (IEA)

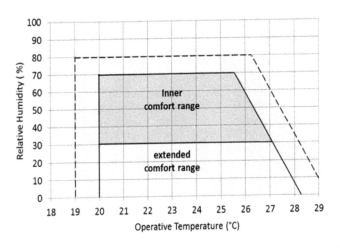

Figure 7.12 Graphical representation of the comfort zone, where humans feel most comfortable in relation to humidity and temperature

7.8.1   Southern European climates

Passivhaus performance has been studied in Southern European climates.[5] The following results were found:

- the use of double glazing is acceptable;
- the use of hygroscopic materials and thermal mass is important;
- adjustable external shading is essential;
- any additional cooling demand ≤ 15 kWh/(m²a);
- it becomes advantageous to use the ground as a heat or cold buffer for tempering the supply air.

Where the required frequency of keeping internal temperatures below 25°C exceeds 10 per cent of the year additional measures are required to manage cooling, such as cross-ventilation and night purge ventilation. If these are not possible, an additional 15 kWh/(m²a) of cooling is permitted and is almost always sufficient because Passivhaus design is very effective at reducing unwanted heat gains.

7.8.2   Southern Hemisphere

The main difference in the Southern Hemisphere for Passivhaus design is to reverse north- and south-facing measures. Dependent upon the latitude and climate, external shading may be needed on the north- and west-facing façades to prevent overheating. A tool is available from Passipedia (http://passipedia.org/), an online knowledge base from the International Passive House Association (iPHA), which allows Southern Hemisphere climate data to be correctly entered in to the PHPP model. Passivhaus buildings are beginning to appear in Australia, southern Africa and South America.

7.8.3   MENA countries

Passivhaus in the MENA (Middle East and North Africa) countries is, at the time of writing, in an early phase. In 2013, to test the concept, two villas in Qatar were constructed side-by-side (see Figure 7.13), one according to the Passivhaus standard and the other according to conventional construction practices in the country. They were then

Figure 7.13 The Qatar Passivhaus villa (with roof covered in solar panels) next to the control villa

monitored. The results were not quite perfect but did indicate that the Passivhaus achieved the Passivhaus comfort criteria, where indoor temperatures were maintained below the 25°C limit, and performed much better than the standard villa. The PV panels on the roof generated sufficient electricity to power the remaining cooling load. The test concluded that the Passivhaus technique is applicable to the hot, arid climate of the MENA region.[6]

Another certified Passivhaus in the UAE has been studied and found to maintain an internal temperature of 22°C to 25°C in all seasons and times, without the need for any traditional cooling device, with a reduced energy consumption of 75 per cent.

7.8.4 Eastern Europe

Bavaria: the residential and commercial building in Kaufbeuren first fulfilled the criteria for the certificate 'Passivhaus Premium'. With a heating demand of only 8 kWh/m²a it has a unique energy efficiency.

At the same time, renewable energy is generated with a 250m² photovoltaic system on the roof.

7.8.5 USA

In the USA, Passivhaus is certified by Passive House Institute US (PHIUS), a non-profit organisation. Certified and Pre-Certified projects at the time of writing totalled more than 1 million square feet across 1,200 units nationwide. The PHIUS+2015 Passive Building Standard, released in 2015, is spurring new growth in passive buildings and is based upon climate-specific comfort and performance criteria. Passivhaus is being developed in California, including one project to date as far south as Los Angeles. Projects are recorded on the website passivehousecal.org

New York City Council passed a law in 2016 prescribing that most new city buildings and major retrofits will need to achieve LEED Gold Standard and cut energy use in half, and mandated that, from 2019, energy performance design targets are to rise to passive house-like levels by 2022.

The Manhattan Borough Board passed a resolution, also in 2016, supporting the investigation and implementation of the Passive House Classic and Passive House Plus (net zero) and Passive House Premium (net positive) standards for potential application to new construction and renovation. It is leading by example with city-owned new buildings and substantial renovations as proofs of concept.

The Board conducted a cost analysis and found that the capital cost difference between a traditional New York City tall building and passive house standard was just 2.4 per cent.

New York Housing Preservation and Development issued in 2016 a request for proposals for the design and construction of what will be the largest passive house certified project in the USA. It will contain at least 400 new affordable dwellings for mixed-income and mixed use in East Harlem.

In addition in Rockaway, New York, a 101-unit passive house affordable apartment building is planned and an existing tall residential apartment block is also being renovated to meet passive house standard using existing products and methodologies.

Figure 7.14 Cornell Tech (latitude 40.7°N) is the world's tallest passive house and one of the first high-rise residential buildings in the world built to passive house standard. Certified under Passive House International (PHI), utilising the PHPP modelling tool. Completed in 2017, it is a 270ft tall tower with 350 residential units and forms a template for green design in New York City and the USA. The tower is scheduled to use 60–70 per cent less energy than a similar construction with a conventional design. Envelope design incorporates thermal bridge-free construction, air tightness to meet .6 ACH, and high R-values/low U-values. This robust approach leads to drastically low heating and cooling loads along with outstanding user comfort and increased durability of the enclosure. The architects, Handel Architects LLP, and developers, The Hudson Companies & The Related Companies, regard it as 'a new paradigm for affordable, high performance buildings'. Passive House consulting by Lois Arena of Steven Winter Association

Source: Handel Architects

Notes

1 The EnerPHit Standard is a good practice refurbishment guide for Passivhaus renovations. See: http://bit.ly/2x3nYzk'
2 Psi (ψ) value: a measurement of heat loss through a given length of a material given in SI units watts per metre Kelvin (W/mK) or in British thermal units per foot per hour per degree Fahrenheit (Btu(th) ·ft·h^{-1}·$^{\circ}$F^{-1})
3 Available from the Passivhaus Institute at http://bit.ly/1zrk2r2
4 Adapted from McLeod, R., Mead, K., and Standen, M., *Passivhaus Primer: Designer's Guide – A Guide for the Design Team and Local Authorities*, BRE Trust and International Passive House Association, Darmsradt, 2015, available at http://bit.ly/Q8aqNM
5 Ibid.
6 See Note 2.
7 'Thermal Comfort Analysis for the First Passivhaus Project in Qatar', Khalfan, May and Sharples, S., Proceedings of SBE16 Dubai, 17–19 January 2016, Dubai–UAE. Available at: https://livrepository.liverpool.ac.uk/2047759/

8 Natural and augmented lighting

Daylight should become the primary light source in buildings for the purposes of promoting health, productivity and sustainability. In the healthcare sector, for example, it has been established the use of daylighting obtains the following benefits:[1]

- sunlight has disinfectant qualities;
- a reduction in the average length of hospital stay;
- quicker post-operative recovery;
- reduced requirements for pain relief;
- quicker recovery from depressive illness;
- benefits with obesity and heart disease.

There are legislative requirements for ensuring adequate daylight provision in new buildings in many countries.

8.1 Light

Light intensity as perceived by the human eye is measured in units called lumens. The perceived effect of illumination is measured in units called lux. One lux (symbol: lx) is equal to the perceived illumination of one lumen per square metre. (In non-SI units, one footcandle is equal to approximately 10 lux.) Note that these units relate to the effect on the human eye, and are not in themselves energy units.

Table 8.1 The number of lux needed for different applications

Lux level	Area or activity
20–30	Car parks, roadways
<100	Corridors, stores and warehouses, changing rooms and rest areas, bedrooms, bars
150	Stairs, escalators, loading bays
200	Washrooms, foyers, lounges, archives, dining rooms, assembly halls and plant rooms
300	Background lighting, e.g. IT office, packing, assembly (basic), filing, retail background, classrooms, assembly halls, foyers, gymnasium and swimming pools, general industry, working areas in warehouses
500	General lighting, e.g. offices, laboratories, retail stores and supermarkets, counter areas, meeting rooms, general manufacturing, kitchens and lecture halls
750	Detailed lighting, e.g. manufacturing and assembly (detail), paint spraying and inspection
1,000	Precision lighting, e.g. precision manufacturing, quality control, examination rooms
1,500	Fine precision lighting, e.g. jewellery, watch making, electronics and fine working

Source: Carbon Trust and lighting manufacturer Veelite

8.1.1   *Daylight factor*

The daylight factor (DF) is the ratio of internal light level to external light level and is defined as follows:

$$DF = (E_i / E_o) \times 100\% \tag{1}$$

where:

E_i = illuminance due to daylight at a point on the indoors working plane

E_o = simultaneous outdoor illuminance on a horizontal plane from an unobstructed hemisphere of overcast sky.

Factors affecting E_i are:

• the sky component (SC): direct light from a patch of sky visible at the point considered;

- the externally reflected component (ERC): the light reflected from an exterior surface and then reaching the point considered;
- the internally reflected component (IRC): the light entering through glazing and reflected from an internal surface.

The illuminance level (lux) at any point being considered is the sum of these:

$$Lux = SC + ERC + IRC \qquad [2]$$

Each of these components may be adjusted by design (e.g. the reflectivity of internal surfaces).

A DF of the following level has these properties:[2]

- <2: not adequately lit – artificial lighting will be required;
- 2–5: adequately lit but artificial lighting may be in use for part of the time;
- >5: well lit – artificial lighting not required except at dawn and dusk. However, glare and solar gain may cause problems.

Light intensity decreases by the square of the distance from the point source. Therefore, 500 lux directed over 10 square metres will be dimmer than the same amount spread over 1 square metre.

Once the lux level has been decided upon and the daylight factor calculated for a particular point, then the daylight autonomy (DA) can be calculated. DA is the percentage of the occupied times of the year (in hours) when the illuminance requirement is met by daylight alone. But this approach has limitations: DF and DA do not say anything about the quality of the daylight, which is subjective. For example, the degree of shading or contrast can affect what may be visually discerned. This may depend upon whether the sky outside is cloudy or clear.

A refinement is climate-based daylight modelling which, at the time of writing, is yet a new art, but software tools are reported to be imminent. This provides a basis for considering daylight holistically, based on building location and façade orientation. It can be integrated with thermal modelling. A suite of daylight metrics is being developed which will use climate-based simulation to meet the visual needs of occupants. The Illuminating Engineering Society (IES) has already formally published two metrics: Spatial Daylight Autonomy

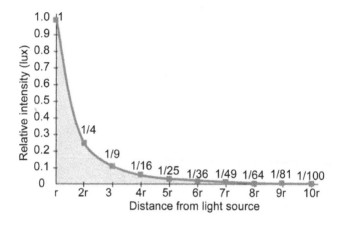

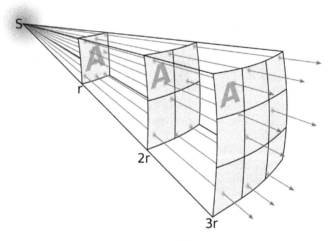

Figure 8.1 Illumination decreases by the inverse square law with distance from the light source. As an example, if a bulb gives off 400 lux at 1m, at a distance of 4m the irradiation will be one 16th of this, or 25 lux (reading from the graph). If 300 lux is required at 4m distance, then 12 lamps each giving off 400 lux would be required (300/25). This illustrates the importance of positioning in lighting

Source: Author and Wiki Commons, author: Borb

and Annual Sunlight Exposure. In the USA, LEED credits are available if LM-83 metrics/modelling is used. Readers are directed to the European Committee for Standardisation, the IES Daylight Metrics Committee and the International Commission an Illumination D3/D6 Technical Committee for further information.

8.2 Techniques for using natural light

Passive daylighting is a system of collecting sunlight to maximise its benefits for lighting, in a controlled manner to avoid unwanted glare. The following tactics may be deployed:

- altering window size, shape, position and orientation;
- glazing coatings;
- reveal angles;
- shading devices (interior and exterior);
- light shelves;
- skylights and rooflights;
- atrium spaces;
- light wells;
- fibre-optic cable networks connected to rooftop light traps;
- tubular daylight devices (sun pipes);
- reflective or pale painted surfaces and interior decor;
- daylight responsive electric lighting controls;
- ceilings might be sloped to direct more light inward.

It is vital to control direct daylight reaching critical visual task areas, so it needs to be filtered. Artificial light (and the use of sun pipes if possible) should be brought in gradually further within spaces, so that there is not a sudden contrast between natural and artificially lit areas. The intention is to direct low-angled daylight high into a space (to reduce the likelihood of excessive brightness).

8.2.1 *Window size, position and latitude*

To capture the requisite amount of solar light, windows will tend to become larger and equator-facing the further towards the poles the building is located. Conversely, the lower the latitude, the smaller they will tend to be, especially on the east and west sides, where the sun's inclination is less in the morning and evening.

8.2.2   Separation from solar gain

The management of daylighting will help to separate it from that of solar heat gain, which may not always be required. This can be achieved by choosing the light-to-solar gain (LSG) of glazing. This is the ratio between the solar heat gain coefficient (SHGC) and visible transmittance (VT). It provides a gauge of the relative efficiency of different glass or glazing types in transmitting daylight while blocking heat gains. The infrared components of the spectrum are absorbed in the glazing. The larger the number, the more light is transmitted without adding excessive amounts of heat.

- *Visible transmittance (VT)* is a fraction of the visible spectrum of sunlight (380 to 720 nanometres), weighted by the sensitivity of the human eye that is transmitted through the glazing of a window, door, or skylight. VT is expressed as a number between 0 and 1, with 0 signifying completely opaque and 1 completely transparent.
- *The solar heat gain coefficient (SHGC)* of glazing (called the 'G-value' in Europe) is the proportion of total solar energy that enters via the window. See section 4.2 for more information.

8.3   Daylighting in hotter climates

The following aspects of solar architecture, discussed in Chapters 4–6, in passive heating/cooling are also of relevance to passive daylighting:

- external shading calculated to keep windows in the shade during summer months;
- making the building wider on an east–west axis than on the north–south axis to minimise the solar gain that enters beneath the overhangs in the morning and afternoon;
- having no skylights and roof windows;
- keeping shutters and curtains closed in the day and open at night. External shades are about 35 per cent more effective than internal ones;
- window overhangs sized relative to the latitude, location and window size. External shading provided by trees; deciduous trees shed their leaves in the winter, allowing the sunlight into the building;
- fixed architectural elements for shading include, besides over-hangs: pagodas, vertical fins, balconies and false roofs;

- internally: light shelves and louvres;
- adjustable elements include: awnings, shutters, blinds, rollers and curtains.

8.4   Daylighting in cooler climates

In cooler climates, many of the above principles apply but they are applied differently:

- equator-facing windows on the east–west axis will permit a large amount of glazing to be used in relation to the building's volume;
- skylights and rooflights can be beneficial as long as overheating in the daytime and heat loss at night and in cold periods are avoided;
- automatically adjusting motorised window-shading-and-insulation devices controlled by sensors that monitor light levels and room occupancy.

8.5   Skylights

Skylights may be either passive or active. Passive skylights simply let daylight enter through glazing in the roof. Active skylights contain a mirror system that tracks the sun across the sky to reflect and direct it where needed.

Optionally, systems are available which reduce daylight during summer months and assist with cooling and ventilation.

8.6   Light shelves

A light shelf is an architectural element that allows daylight to penetrate deep into a building (see Figures 8.2, overleaf and 8.5, page 162). They are effective on equator-facing façades, but often ineffective on east or west elevations of buildings. A horizontal, light-reflecting overhang is placed above eye-level and has a high-reflectance upper surface to reflect daylight onto the ceiling. They are commonly made of an extruded aluminium chassis system and aluminium composite panel surfaces.

Light shelves and louvred systems make it possible for daylight to penetrate the space up to four times the distance between the floor and the top of the window. They are generally not used in tropical or desert climates due to the intense solar heat gain.

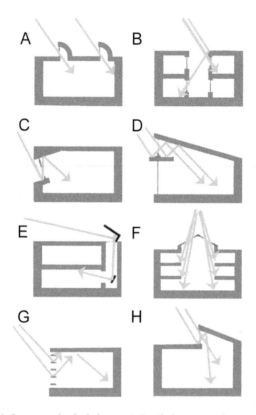

Figure 8.2 Strategies for daylighting: A: Rooflights oriented equator-wards; B: Central courtyard; C: Recessed and down-angled window, with reflectors; E: Split window with light shelf; F: Atrium; G: Light shelves in window; H: Oriented rooflight with reflectors

Source: Author

Exterior louvre systems are vertical or horizontal fins whose opening and shutting, rather like venetian blinds, can be controlled automatically (See Figure 8.3). They are used to prevent too much sunlight from reaching a window. They can also have reflective surfaces to direct light inside while avoiding glare.

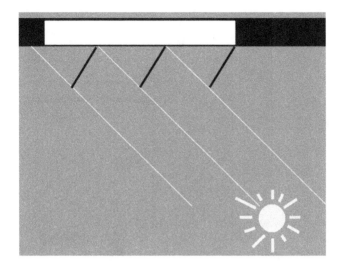

Figure 8.3 Plan view of a window in a wall showing adjustable vertical louvre positioning on the outside to prevent glare and/or reflect deeper light into a building

Source: Author

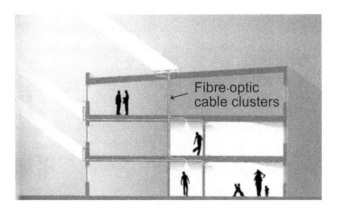

Fibre-optic
cable clusters

Figure 8.4 Light collector on the roof with clusters of fibre-optic cables conducting the light into deep rooms below

Source: Author

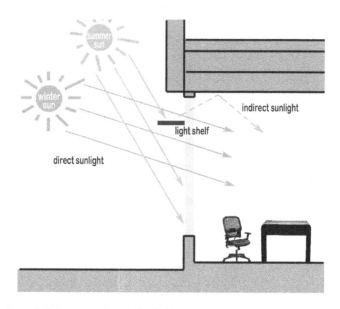

Figure 8.5 Principle of the light shelf. It may be inside or outside, fixed or adjustable

Source: Author

8.7 Sun pipes

Also called light tubes or daylight pipes, sun pipes have reflective inner surfaces to take light from a roof level deep into lower floors (see Figure 8.6). Compared to conventional skylights and other windows, they offer better heat insulation properties and more flexibility for use in inner rooms, but less visual contact with the external environment. Light can be brought into spaces also by using atria and light guidance systems.

8.8 Operable shading and insulation devices

A design with too much sun-facing glass can result in excessive heating, or uncomfortably bright living spaces at certain times of the year, and excessive heat transfer on winter nights and summer days.

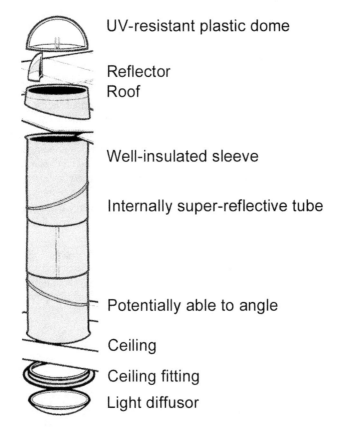
UV-resistant plastic dome

Reflector
Roof

Well-insulated sleeve

Internally super-reflective tube

Potentially able to angle

Ceiling

Ceiling fitting

Light diffusor

Figure 8.6 Exploded diagram of a sun pipe or light well

Source: Author

Although the sun is at the same altitude six weeks before and after the solstice, the heating and cooling requirements before and after the solstice are significantly different. Variable cloud cover influences solar gain potential. This means that latitude-specific fixed window overhangs, while important, are not a complete seasonal solar gain control solution.

Control mechanisms include:

- manual-or-motorised interior insulated drapes;
- shutters;
- exterior roll-down shade screens;
- retractable awnings.

These can compensate for differences caused by thermal lag or cloud cover, and help control daily/hourly solar gain requirement variations. Automated systems monitor temperature, sunlight, time of day and room occupancy. These can precisely control motorised window-shading-and-insulation devices. In large office buildings, facilities and retail centres, a successful lighting design that incorporates any of these architectural features would be integrated with the artificial lighting system using advanced lighting controls, and with the building energy management system (BEM) if there is one. Three types of controls are available:

1. those that turn lights off when there is ample daylight;
2. control of individual lamps in zoned circuits that are progressively further from windows, to provide intermediate levels of light;
3. dimming controls which continuously modulate the power to lamps to complement the amount of daylight.

8.9 Satisfying the remaining lighting requirements artificially

The efficacy of a lamp is measured in the number of lumens it produces per watt input. The strategy is to maximise the number of lumens obtainable for the least number of watts. If an office that requires 500 lux has an area of 100 square metres and the lamps are 2 metres above the desk level, then this will need 500 × 100 × 4 = 200,000 lumens. This office could therefore be lit (at night) by (using the figures in Table 8.2): 200,000/70 = 2156W of CFLs, 200,000/90 = 2,224W of LEDs, or 2,000W of high-frequency fluorescent tubes.

8.9.1 *Light-emitting diodes (LEDs)*

For satisfying almost all indoor uses, types of LED lamps are now available and come in a full range of colour rendering and

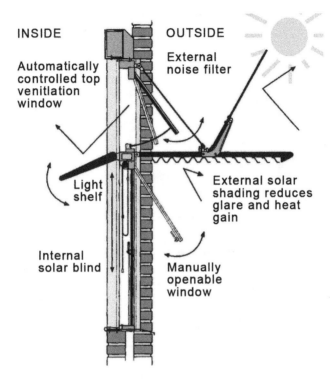

INSIDE

Automatically
controlled top
venitlation
window

Light
shelf

Internal
solar blind

OUTSIDE

External
noise filter

External solar
shading reduces
glare and heat
gain

Manually
openable
window

Figure 8.7 Cross-section illustrating various types of shading and light-directing fixtures for and around sun-facing windows to maximise daylighting for interiors while controlling glare around the year and through the day

Source: Author, adapted from Creative Commons image created by Mr7uj

temperatures. They can last up to 2.5 times longer than CFLs and 25 times longer than incandescents, while consuming much less power. Buy with care as not all LED lamps are of the same quality.

8.9.2   *Luminaires*

A luminaire is the fixture that holds the lamp. There are many types but all contain five main components: the housing, control gear, lamp

Table 8.2 The performance of typical 12V lamps

Lamp type	Rated watts (W)	Light output lumens (lm)	Efficacy (lm/W)	Lifetime (hours)
Incandescent globe	15	135	9	1,000
Incandescent globe	25	225	9	1,000
Halogen globe	20	350	18	2,000
Batten-type fluorescent (with ballast)	6	240	40	5,000
Batten-type fluorescent (with ballast)	8	340	42	5,000
Batten-type fluorescent (with ballast)	13	715	55	5,000
PL-type fluorescent (with ballast)	7	315	45	10,000
LED lamp*	3	180	30-100	>50,000

Source: Manufacturers' data
Note: *The performance of LEDs varies considerably according to the manufacturer. Choosing the right LED products is very important

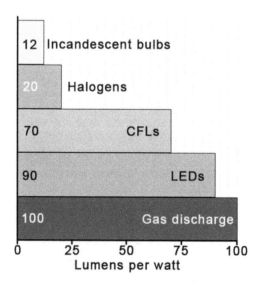

Figure 8.8 The average efficacy of different types of lighting
Source: Author

holder, lamp and reflector. Reflectors are essential to direct light, maximise its usage and, thus, reduce the quantity needed. Some luminaires may also include a diffuser.

Their efficiency is measured by their light output ratio (LOR). The higher this is, the better. An important characteristic for efficiency purposes is the reflectiveness of the material used. Satin chrome reflects only half the light, whereas aluminium coated with silver reflects 90 per cent of it. Luminaires in offices should be designed to avoid glare onto screens. This type should not be used in retail environments, which have different requirements.

8.9.3   *Lighting controls*

Control gear is used to preserve lamp life for gas discharge lighting and can additionally provide manual or automatic dimming and switching. Modern controls use high-frequency electronic control equipment; this produces more light with less power. However, not all high-frequency lighting can be dimmed.

* *Automatic and manual controls* should be combined in areas where people need to control their own level of lighting. This means that users should be able to dim and switch off and on lighting as they require, but it may also be switched off or dimmed automatically when not needed.
* *Dimming* gives occupants more control over their light levels and still yields savings.
* *Adjustable light level sensors* can automatically turn a light on and off in response to changing amounts of daylight. These can be used outside, such as in car parks or streets, or inside, to maintain an even level of light as the light from outside changes during the day.
* *Occupancy sensors* can tell whether a space is occupied and control lighting accordingly. They are appropriate for internal or external lighting and come in several types:

    1. *Doppler sensors* work by sending out high-frequency sound waves and listening for the bounce back. When it is returned at a different frequency it knows there is a moving object around. It then sends a signal to the dimmable ballast to raise the light levels. When no movement is detected after

a certain period of time, the light will return to its original level.

2. *Passive infrared (PIR)* motion sensors are used in many cases and also control flyers and taps, but for lighting can be vulnerable to dust and blocking objects, can be confused by radiators and fires and can have a shorter lifespan.

3. In some circumstances the use of a *manual switch with a timer* built into it may be appropriate.

Most control systems need additional connectivity. Nowadays, wireless technology is often cheaper to install than wired connections. Digital addressable lighting interfaces (DALI) is one standard used for intelligent lighting management, globally. It is a protocol set out in IEC 60929 and IEC 62386, which are technical standards for network-based systems that control lighting in buildings. A DALI network consists of a controller and one or more lighting devices (e.g. electrical ballasts and dimmers). Each device is assigned a unique static address, allowing it to be remotely controlled. DALI also attempts to reduce the standby parasitic power losses of control equipment.

Notes

1   Source: www.glassforEurope.com/images/cont/225_12633_file.pdf
2   CIBSE Lighting Guide 10: Daylighting and window design, Year: 1999, ISBN 0-900953-98-5, CIBSE.

9 Notes on technology and building occupation

A 'passive' solar-conscious building is so called because the dominating design principle is to maximise the use and benefits of sunshine by means of the intrinsic design and orientation of the building. Having designed such a building this does not mean that 'active' aspects, especially relating to control, are not incorporated. 'Building management systems' are examples of such active support. However, equally 'active' are the occupants when, for instance, they open windows or close shutters. The essence of all such activities is that artificial energy supplies are minimised and individuals have freedom to enjoy their surroundings as they wish.

9.1 Building management systems (BMSs)

Large, well-integrated buildings incorporate a BMS – a network of controls connected independently or centrally, operated automatically or manually, that measure and respond to both external and internal stimuli. Increasingly, these operate on a local area wi-fi network.

- Basic functions include monitoring, controlling and optimising energy usage.
- Advanced systems can measure rainfall, wind speed and occupancy density.

The aim is to manage the building environment so that the supply of energy perfectly balances with the use made of it.

The level of control may be refined – for example, by splitting areas into different zones within, say, an open-plan office, warehouse, large retail unit or factory. Systems are available that extend the management of resources from energy to include water, air and steam. So-called WAGES (water, air, gas, electrical, steam) systems are intended for industrial applications. Increasingly, software can be cloud-based and involve modular, downloadable apps for many different purposes.

A hybrid ventilation system, for example, can be managed in three ways:

- *active*: will automatically respond to environmental conditions; can be interrogated and adjusted by the energy manager;
- *informed occupancy*: gives alerts to staff to make necessary changes, e.g. draw and pull back shades daily, control ventilation and lighting;
- *seasonal lockout*: the custodial staff make changes such as closing vents at the change of the seasons.

Even when systems are fully automatic, the facility for occupants to control their environment is important – for example, the ability to override and adjust lighting and to control shading louvres.

9.2   Drawbacks of MVHR, BMS and other technology

Experience shows that the more complex a system, the more likely it is to:

1. break down (especially the electronic components), and so have increased life-cycle costs;
2. not be used correctly by occupants, so occupants or the facility/ energy manager will need regular training about how not to interfere with the system's proper function and in its correct use;
3. have an increased ecological footprint and embodied energy, so is important to make sure that the life-cycle cost of running the BMS, including maintenance, is not greater than the benefits.

There is, therefore, great virtue in simplicity. Some passive building designers eschew MVHR on these grounds, especially in small and domestic buildings. They argue that although, on paper, there may be benefits in introducing the technology, in practice these benefits are marginal. Therefore, the added expense and the embodied energy in the technology are unnecessary.

9.3   Building information modelling software (BIMS)

For large projects, building information modelling is used at the design stage. This is collaborative software for sharing knowledge and modelling the functioning and energy use of a building through its entire life-cycle.

Green Building XML (gbXML) is an emerging schema, a subset of the BIM efforts, focused on green building design and operation. gbXML is used as input in several energy simulation engines – see: http://gbxml.org/

While BIMs are certainly useful, they must be deployed with the caveat that buildings are frequently not constructed according to their original design, so there needs to be close cooperation between the building designer, the project managers and the builders for the predicted savings to be ensured.

Following on, occupants need to be educated in the building use. An easy-to-use and disseminate manual should be provided.

10 Resources

The International Energy Agency's (IEA) Solar Heating and Cooling Programme (SHC), running since 1977, promotes the use of all aspects of solar thermal energy and has many resources, available at: http://www.iea-shc.org/

10.1 Software tools

IEA-SHC's Task 41: Solar Energy and Architecture gathered researchers and practising architects from 14 countries in a three-year project whose aim was to identify the obstacles architects are facing when incorporating solar design in their projects. These resulting publications are of interest:

- *Case Studies of Passive Solar Buildings*: http://task41.iea-shc.org/casestudies/
- *Solar Design of Buildings for Architects: Review of Design Tools*: http://task41.iea-shc.org/data/sites/1/publications/T41B3_approved-Jul12.pdf

The latter does not rank tools against each other, but provides an overview of tools' capabilities. The survey found that for solar design at the early design stage only some of the existing computer-aided architectural design (CAAD) tools and simulation tools can be used by architects. Tools covered include the following.

10.1.1 Simple graphical tools

- Solar charts/sun path diagrams
- Physical models

10.1.2 Digital tools

- AutoCAD (does not support any form of solar modelling)
- ArchiCAD
- Google Sketchup: The OpenStudio (SketchUp plugin with EnergyPlus)
- Autodesk® Revit® Architecture 2012
- VectorWorks
- Autodesk® Ecotect®
- Autodesk® Project Vasari 2.1
- RETScreen® International
- Radiance (a suite of software for the analysis and visualisation of lighting in design)
- IES VE
- SolarBILANZ
- bSol
- DAYSIM
- Design Performance Viewer (DPV)
- Lesosai
- Polysun
- PVsyst
- PV*Sol Expert 5.0
- T*Sol Pro 5.0

PHPP was not evaluated in that survey. See section 7.5 for details of PHPP. In a survey of architects, Ecotect was found to be the most popular. Also of interest was Google Sketchup, widely used by most Schools of Architecture worldwide. This 3D CAD program has a very friendly interface and is easy to learn. The program reads all dxf and dwg files and a free version is available on the internet. The US Department of Energy has released a free Sketchup plugin version of the Energy Plus simulation program. Energy Plus is the most used simulation program all around the world and is free software. The Daysim software is also free software that gives interesting outputs such as daylight autonomy.

OpenStudio (openstudio.net) is a cross-platform collection of software tools to support whole building energy modelling using EnergyPlus and advanced daylight analysis using Radiance. Highlights of OpenStudio plugin include the ability to:

- create, edit and view EnergyPlus input files within SketchUp;
- add internal gains and simple outdoor air (i.e. natural ventilation) for load calculations;

- add the ideal HVAC system for load calculations;
- set and change default constructions;
- add daylighting controls and illuminance map.

Details can be added according to the available level of details of the model. The results are obtained as spreadsheets or text output files. The program can perform the following analyses:

1. solar radiation incident, absorbed and transmitted;
2. advanced shading (including scheduled controls);
3. solar heat gain from windows;
4. advanced daylighting.

Constructions can be detailed layer by layer, or can be selected from an existing database.

Paid-for software, however, often has greater functionality, including the option to run cost-benefit analyses for different options, calculate CO_2 emission savings, fuel consumption and other quantitative values.

Note: Software should not be relied upon absolutely. Calculations made by software should always be checked against real-world measurements or hand calculations so as to avoid errors – which do occur, sometimes yielding absurd results. The use of advanced simulation packages can so easily lead to incorrect results. The user should always question whether the result relates to their personal experience and understanding.

10.1.3 Software packages for natural ventilation analysis

- FloVENT: calculates airflow, heat transfer and contamination distribution for built environments using computational fluid dynamics. https://www.mentor.com/
- FLUENT: a computational fluid dynamics program useful in modelling natural ventilation in buildings, calculation of air flow modelling, contaminant transport, room air distribution, temperature and humidity distribution and thermal comfort. It models air flow under specified conditions, so additional analysis is required to estimate annual energy savings. http://www.ansys.com/

- UrbaWind: models the wind in urban area and calculates automatically the natural air flow rate in the buildings, according to the surrounding buildings effects and the local climatology. Incorporates a natural ventilation module that estimates natural air ventilation inside buildings to design bioclimatic, passive and high-performance buildings. http://meteodyn.com/
- DOE-2: a comprehensive hour-by-hour simulation; daylighting and glare calculations integrate with hourly energy simulation (freeware). http://doe2.com/
- EnergyPlus: a whole building energy simulation program to model both energy consumption – for heating, cooling, ventilation, lighting and plug and process loads – and water use in buildings. https://energyplus.net/
- DELPHIN is a simulation program for the coupled heat, moisture and matter transport in porous building materials. The DELPHIN software is used for calculation of thermal bridges, including evaluation of hygrothermal problem areas (surface condensation, interstitial condensation), design and evaluation of inside insulation systems, evaluation of ventilated façade systems, ventilated roofs, transient calculation of annual heating energy demand (under consideration of moisture dependent thermal conductivity), drying problems (basements, construction moisture, flood) and calculation of mould growth risks. http://bauklimatik-dresden.de/
- The CBE Thermal Comfort Tool to evaluate thermal comfort. Users provide operative temperature (or air temperature and mean radiant temperature), air speed, humidity, metabolic rate, and clothing insulation value, the tool evaluates predicted thermal sensation on a scale from −3 (cold) to +3 (hot). Compliance is achieved if the conditions provide thermal neutrality, measured as falling between −0.5 and +0.5 on the PMV scale. Licensed under a Creative Commons License. See: http://bit.ly/2uLarf2
- WUFI® software products allow realistic calculation of the transient coupled one- and two-dimensional heat and moisture transport in walls and other multi-layer building components exposed to natural weather. WUFI® is an acronym for Wärme Und Feuchte Instationär which, translated, means heat and moisture transiency. It addresses vapour diffusion and moisture transport in building materials. The software has been validated by detailed comparison with measurements obtained in the laboratory and in real-world tests. Versions are available for many countries around the world. https://wufi.de/en/

- WUFI® Passive is a user-friendly passive building energy modelling software tool. Versions exist for different territories and climate zones: North America and Europe. It combines passive building energy modelling with hygrothermal analysis for preventing moisture issues and optimising for cooling/dehumidification strategies. Toggles between metric and imperial units, and is compatible with Google SketchUp. It is of help with the design process for Certified Passive House Consultants. A free version is available.
- Not software, but instructions for designing a cool tower. http://bit.ly/2huyAjg

10.2   Data sources and calculators

10.2.1   *Buildings*

- Full U-value calculation conventions are available in BR443:2006 for free. http://bit.ly/1o5Lcyd
- Thermal conductivity of materials tables. http://www.kayelaby.npl.co.uk/general_physics/2_3/2_3_7.html and http://www.engineeringtoolbox.com/thermal-conductivity-d_429.html
- U-value calculator. http://www.vesma.com/tutorial/uvalue01/uvalue01.htm
- Online Heat Loss Calculator. http://www.engineeringtoolbox.com/heat-loss-transmission-d_748.html
- A free tool for calculating the thermal properties of construction elements is available from http://bit.ly/2s1efYm
- The ASHRAE Handbook: a checklist of procedures, design data, recent industry practices, indoor air and environmental quality, building design, US building codes and standards, building systems and controls and specific application guidance for mechanical engineering, engineered systems for buildings or architecture. https://www.ashrae.org/resources–publications/bookstore/handbook-online
- The ASHRAE Advanced Energy Design Guides (AEDG) provide recommendations for achieving energy savings over the minimum code requirements of ANSI/ASHRAE/IESNA Standard 90.1 for non-residential buildings. They are free to download after registering on the site. Building typologies include retail, offices, grocery stores, K-12 schools, warehouses/storage, lodging and healthcare facilities

- The NetZEB Evaluation Tool – an Excel-based platform tool that enables energy balance, operating cost and load match index calculation for predefined selected definitions of net-zero energy buildings. It aims at evaluating solutions adopted in building design with respect to different NetZEB definitions (for building designers), assessing the balance in monitored buildings (for energy managers) and assisting the upcoming implementation process of NetZEBs within the national normative framework (for decision-makers). http://task40.iea-shc.org/net-zeb
- Carbon calculator for the embodied carbon of buildings. http://www. woodworks.org/design-and-tools/design-tools/online-calculators/
- Radiance is a suite of tools for performing lighting simulation by using ray tracing, but it's necessary also to take into account obstruction caused by wall hangings or furniture. http://www. radiance-online.org/
- Passive House Institute USA. http://www.phius.org/

10.2.2   Climate data

Climate data are based either on ground-based measurements or satellite data. Data are often interpolated where no local measurements exist, or used as representative of a surrounding zone with similar climatic conditions. The following is a list of examples of worldwide climate data sources:

- NOAA Sunrise/Sunset and Solar Position Calculators and spreadsheets. http://1.usa.gov/1g8xFme
- NREL data. www.nrel.gov/gis/data_solar.html and www.nrel.gov/rredc/pvwatts
- Calendar of sunrise, sunset, noon daylight at any location for an entire year. The table shows the time and azimuth in degrees. http://bit.ly/1njenNl
- Sun position calculator producing sun path diagrams from location found using Google maps, rendering elevation, azimuth for any times, latitude, longitude, etc. http://bit.ly/1kM2aM0
- Sun path diagrams for each 1° of latitude for the Northern and Southern Hemisphere. http://bit.ly/1mCSJQv
- PVGIS: solar radiation data for Europe, Africa and South-West Asia, and ambient temperature for Europe, plus terrain and land cover. http://re.jrc.ec.europa.eu/pvgis

- Solar path calculator; a solar spectrum calculator; an installed system cost calculator; and calculators for solar cell operation and future module prices. http://www.pvlighthouse.com.au/calculators/calculators.aspx
- Degree days: degreedays.net. US data: Climate Predication Center: www.cpc.noaa.gov/products/monitoring_and_data; Canadian Integrated Mapping and Assessment Project: http://bit.ly/2sQG6wU
- NASA cloud cover data. http://bit.ly/1GZ2UYy
- NASA surface meteorology and solar energy (SSE) database of solar insolation, rain and wind data. http://eosweb.larc.nasa.gov/sse/
- TMY files: typical annual profiles of exterior climate data such as ambient temperatures, wind direction and velocity, precipitation, direct and diffuse irradiance, (free) hourly climate data for many locations worldwide in the epw format. https://energyplus.net/weather/sources
- World solar irradiation data: detailed maps, and maps of direct normal irradiation (free). http://solargis.info/doc/71
- Irradiation data by country. www.sealite.com.au/technical/solar_chart.php The unit used is $kWhm^{-2} day^{-1}$
- Many weather stations are listed with individual station IDs at the World Meteorological Organization (WMO) – for example, through the World Data Center for Meteorology and their publication of World Weather Records (WWR). Weather stations are usually maintained by the national meteorological organisations. These organisations therefore are a first point of contact when searching for climate data
- The software Meteonorm (www.meteonorm.com) is a worldwide meteorological database. Climate data can be extracted in a variety of formats (including PHPP) for any location on the globe and different reference periods. It is used by PHIUS.
- Worldwide data from the NASA Langley Research Center Atmospheric Sciences Data Center POWER Project has been converted by the Passive House Institute into the format required for the PHPP and is available for download from the climate data tool integrated into Passipedia (www.passipedia.com).
- The American Society of Heating, Refrigerating and Air-Conditioning Engineers (ASHRAE) publishes a selection of hourly climatic datasets named International Weather for Energy Calculations (IWEC).

- A number of climate data sets from a variety of sources have been converted into the data format required for the simulations software EnergyPlus and is available on the website of the US Department of Energy, Department of Energy Efficiency and Renewable Energy.
- Note: The PHPP contains monthly data in the required format for a large number of international locations from various original data sources.

10.3   Standards

- ASHRAE 41.2-1987 Standard Methods for Air Velocity and Airflow Measurement.
- ASHRAE 55 Thermal Environmental Conditions for Human Occupancy.
- ASHRAE 62.1-2013 Ventilation for Acceptable Indoor Air Quality.
- ASHRAE Standard 90.2-2007 Energy Efficient Design of Low-Rise Residential Buildings.
- ASHRAE/IES 90.1-2013 Energy Standard for Buildings Except Low-Rise Residential Buildings.
- BS 5925 Code of practice for ventilation principles and designing for natural ventilation.
- BS EN 1995 Common rules and rules for buildings, structural fire design and bridges.
- California's Title 24 Building Energy Efficiency Standard.
- EN 15251 Indoor environmental input parameters for design and assessment of energy performance of buildings addressing indoor air quality, thermal environment, lighting and acoustics.
- EN 15251 provides a firm framework of IEQ conditions influencing energy performance, as well as principles for design and calculation.
- EnerPHit Standard Certification criteria for refurbished buildings.
- IEC 60929 and IEC 62386: technical standards for network-based systems that control lighting in buildings.
- ISO 10051:1996 Thermal insulation – Moisture effects on heat transfer – Determination of thermal transmissivity of a moist material.
- ISO 11855 Building environment design – Design, dimensioning, installation and control of embedded radiant heating and cooling systems.

- ISO 13612 Heating and cooling systems in buildings – Method for calculation of the system performance and system design for heat pump systems.
- ISO 13790:2008 Energy performance of buildings – Calculation of energy use for space heating and cooling.
- ISO 13791:2012 Thermal performance of buildings – Calculation of internal temperatures of a room in summer without mechanical cooling – General criteria and validation procedures.
- ISO 13792:2012 Thermal performance of buildings – Calculation of internal temperatures of a room in summer without mechanical cooling – Simplified methods.
- ISO 15927 Hygrothermal performance of buildings – Calculation and presentation of climatic data.
- ISO 1984 Moderate Thermal Environments – Determination of the PMV and PPD Indices and Specification of the Conditions for Thermal Comfort.
- ISO 6781:1983 Thermal insulation – Qualitative detection of thermal irregularities in building envelopes – Infrared method.
- ISO 7345:1987 Thermal insulation – Physical quantities and definitions.
- ISO 7726 Ergonomics of the thermal environment – Instruments for measuring physical quantities.
- ISO 9050:2003 Glass in building – Determination of light transmittance, solar direct transmittance, total solar energy transmittance, ultraviolet transmittance and related glazing factors.
- ISO 9251:1987 Thermal insulation – Heat transfer conditions and properties of materials – Vocabulary.
- ISO 9288:1989 Thermal insulation – Heat transfer by radiation – Physical quantities and definitions.
- Passivhaus standard (see www.passiv.de).

10.4   Details

Free architectural drawings are available on the internet for passive house construction.

- The UK's Energy Saving Trust's Enhanced Construction Details for preventing thermal bridging on different construction types are available at: http://bit.ly/2hWjEee
- Other architectural details are available on the foursevenfive website at: http://bit.ly/2hS6YYe

Appendix 1
A ten-step design and build strategy[1]

1. Site selection

   - Secure an optimum location, as free as possible from non-useful shading relative to the seasons and time of day.
   - Research the available solar resource and wind factors for the site using local and freely available data.
   - Orient the design to the elements optimally.

2. Concept development

   - Minimise shade in winter, minimising parapets, projections, non-transparent balcony enclosures, divider walls etc.
   - Choose a compact building structure with low skin-to-volume ratio and without unnecessary recesses. Use opportunities to combine buildings.
   - Survey and model, using software such as PHPP (see Chapter 10) the expected internal and external heat gains and cooling requirements, and other building energy loads.
   - The following energy performance targets and air changes per hour define the Passivhaus standard and must be met in order for certification to be achieved:

     Specific heating demand ≤ 15 kWh/m^2/yr
     Specific cooling demand ≤ 15 kWh/m^2/yr
     Specific heating load ≤ 10 W/m^2
     Specific primary energy demand ≤ 120 kWh/m^2/yr
     Air changes per hour ≤ 0.6 @ 50 Pascals

- Optimise glazing, shading and aspect/form according to latitude and climate zone, to maximise the use of daylighting, balancing against the appropriate heat gains.
- In temperate/hot climate zones concentrate the utility areas in the parts of the building that are coolest in the summer.
- Model and decide on the ventilation scheme making best use of the stack effect and Bernoulli principle. Once you have maximised the opportunities to use natural ventilation techniques then decide whether additional mechanical ventilation (with heat recovery) is needed.
- Thermally separate the basement from ground floor (including cellar staircase), make airtight and thermal bridge free.
- Derive an initial energy use estimate.
- Evaluate the potential for renewable energy technologies: solar thermal, PV, wind, heat pumps, etc. Remember that heat pumps can be reversed to use for cooling.
- If supplementary heating is required consider use of underfloor heating, linked to heat pumps if they are deployed, to save energy (water or electric).
- Check the possibility of government subsidies.
- Commence consultations with the building authority.
- Contract agreement with architects, including a precise description of services to be rendered.

3. Construction plan and building permit planning

- Select the building style – thermally massive or light.
- Sketch out a design concept, floor plan, energy concept for ventilation, cooling, heating and hot water.
- Floor plan: short pipe runs for hot/cold water and sewage.
- Consider the space required for utilities (cooling/heating, ventilation etc.).
- If used, plan for short ventilation ducts: cold air ducts outside, warm ducts inside the insulated building envelope.
- Recalculate and minimise the energy demand, using digital tools.
- Plan the insulating thickness of the building envelope and avoid thermal bridges.
- Calculate cost estimate.
- Negotiate the building project (pre-construction meetings).

4. Final planning of the building structure (detailed design drawings)

   • Insulation of the building envelope: the absolute U-values will vary according to context (location, form etc.), but in general aim for:
     walls, floors and roofs ≤ 0.15 W/m²K;
     complete window installation ≤ 0.85 W/m²K.
   • Design thermal bridge free and airtight connection details.
   • Specify windows that comply with passive house standard: optimise type of glazing, thermally insulated frames, glass area, coating, shading.

5. If used, final planning of ventilation (detailed system drawings)

   • General rule: hire a specialist.
   • Ventilation ducts: short and sound-absorbing. Air flow velocities below 3 m/s.
   • Include measuring and adjusting devices.
   • Take acoustic insulation and fire protection measures into account.
   • Air pathways: avoid air current short-circuiting.
   • Consider the air throws of the air vents.
   • Provide for overflow openings.
   • If heat recovery is used, install in the temperature-controlled area of the building shell.
   • Additional insulation of central and back-up unit may be necessary. Soundproof the devices. Thermal energy recovery rate should be > 80 per cent.
   • Airtight construction to be checked at every stage, yet using breathable materials (see later for explanation).
   • The ventilation system should be user-adjustable.
   • Optional: ground- or water-source heat pump (air or water as medium) and/or air pre-cooling/heating pipes; may be reversible for summer cooling and winter heating.

6. Final planning of the remaining utilities (detailed plumbing and electrical drawings)

   • Plumbing: install short and well-insulated pipes for hot water in the building envelope. For cold water install short pipes insulated against condensation water. Use no greater bore than needed to conserve water and heat. Include separate rainwater collection and delivery piping, labelled.

- Use water-saving fittings.
- Sub-roof vents for line breathing (vent pipes).
- Plumbing and electrical installations: avoid penetration of the airtight building envelope – if not feasible, install adequate insulation.
- Use the most energy-saving appliances/equipment.
- Situate switches for ring mains alongside light switches to enable easy switching off of phantom loads when leaving rooms/building.
- Plan installation of (perhaps wireless) building energy monitoring system.

7. Call for tenders and awarding of contracts

- Plan for quality assurance measures in the contracts.
- Set up a construction schedule.

8. Assurance of quality by the construction supervision

- Thermal bridge free construction: schedule on-site quality control inspections. Take photographs.
- Check of airtightness: all pipes and ducts must be properly sealed, plastered or taped. Electrical cables penetrating the building envelope must be sealed also between cable and conduit. Flush mounting of sockets in plaster and mortar. Take photographs.
- Check of thermal insulation for ventilation ducts and hot water pipes.
- Seal window connections with long-lasting adhesive tapes or plaster rail. Apply interior plaster from the rough floor up to the rough ceiling.
- n50 airtightness test: have a blower door test done during the construction, when the airtight envelope is complete but still accessible – that is, before finishing the interior work, but after completion of the electricians' work (in concert with the other trades), including detection of all leaks.
- Ventilation system: ensure easy accessibility for filter changes. Adjust the air flows in normal operation mode

by measuring and balancing the supply and exhaust air volumes. Balance the supply and exhaust air distribution. Measure the system's electrical power consumption.

- Quality control check of all cooling, heating, plumbing and electrical systems.

9. Final inspection and auditing.
10. If possible conduct post-occupancy monitoring to determine if building performs as expected.

Note

1 With acknowledgements to Isover, St Gobain.

Appendix 2
Units

Many units are explained in context within the text, such as U-values, R-values and perms. The following are not.

SI units

The International System of Units (abbreviated as SI) is the modern form of the metric system, and is the most widely used system of measurement. There are seven basic units:

Unit	Symbol	Application
metre	m	length
kilogram	kg	mass
second	s	time
ampere	A	electric current
kelvin	K	thermodynamic temperature
mole	mol	amount of substance
candela	cd	luminous intensity

Many other units are derived from them:

Name	Symbol	Quantity	Expressed in terms of other SI units	Expressed in terms of SI base units
radian	rad	angle		$m \cdot m^{-1}$
steradian	sr	solid angle		$m^2 \cdot m^{-2}$
newton	N	force, weight		$kg \cdot m \cdot s^{-2}$

Name	Symbol	Quantity	Expressed in terms of other SI units	Expressed in terms of SI base units
pascal	Pa	pressure, stress	N/m^2	$kg \cdot m^{-1} \cdot s^{-2}$
joule	J	energy, work, heat	$N \cdot m$	$kg \cdot m^2 \cdot s^{-2}$
watt	W	power, radiant flux	J/s	$kg \cdot m^2 \cdot s^{-3}$
coulomb	C	electric charge or quantity of electricity		$s \cdot A$
volt	V	voltage (electrical potential difference), electromotive force	W/A	$kg \cdot m^2 \cdot s^{-3} \cdot A^{-1}$
degree Celsius	°C	temperature relative to 273.15 K		K
lumen	lm	luminous flux	$cd \cdot sr$	cd
lux	lx	illuminance	lm/m^2	$m^{-2} \cdot cd$

Other heat units

therm (symbol thm): a non-SI unit of heat energy equal to 100,000 British thermal units (Btu), approximately the equivalent of burning 100 cubic feet (2.83 cubic metres) of natural gas (MBtu = millions of Btus).

toe = tonnes of oil equivalent (Mtoe = millions of toe), a unit of energy defined as the amount of energy released by burning one tonne of crude oil. As different crude oils have different calorific values, the exact value is defined by convention, but several slightly different definitions exist. It is approximately 42 gigajoules. The International Energy Agency defines one tonne of oil equivalent (toe) as follows:

1 toe = 11.63 megawatt-hour (MWh)
1 toe = 41.868 gigajoules (GJ)
1 toe = 10 gigacalorie (Gcal) – using the international steam table calorie (calIT) and not the thermochemical calorie (calth)
1 toe = 39,683,207.2 British thermal unit (Btu)
1 toe = 1.42857143 tonne of coal equivalent (tce)
1 toe = 7.33 barrel of oil equivalent (boe).

Conversion factors

To:	TJ	Gcal	Mtoe	MBtu	GWh
From:	Multiply by:				
terajoules (TJ)	1	238.8	2.388×10^{-5}	947.8	0.2778
gigacalories (Gcal)	4.1868×10^{-3}	1	10^{-7}	3.968	1.163×10^{-3}
million tonnes of oil equivalent (Mtoe)	4.1868×10^{4}	10^{7}	1	3.968×10^{7}	11630
million British thermal units (MBtu)	1.0551×10^{-3}	0.252	2.52×10^{-8}	1	2.931×10^{-4}
gigawatt hours (GWh)	3.6	860	8.6×10^{-5}	3412	1

From	To kWh. Multiply by:
therms	29.31
Btu	2.931×10^{-4}
MJ	0.2778
toe	1.163×10^{-4}
kcal	1.163×10^{-3}

- A unit of measure converter is at: http://bit.ly/1iXDzqP

Example:

Conversion of 100,000 Btu to kWh: 100,000 Btu = 100,000 × 2.931×10^{-4} kWh = 29.31kWh

Conversion factors for mass

To:	kg	T	lt	st	lb
From:	Multiply by:				
kilogram (kg)	1	0.001	$9.84 \times 10_{-4}$	1.102×10^{-3}	2.2046
tonne (t)	1000	1	0.984	1.1023	2204.6
long ton (lt)	1016	1.016	1	1.120	2240.0

To:	kg	T	lt	st	lb
From:	Multiply by:				
short ton (st)	907.2	0.9072	0.893	1	2000.0
pound (lb)	0.454	4.54×10^{-4}	4.46×10^{-4}	5.0×10^{-4}	1

Conversion factors for volume

To:	gal US	gal UK	bbl	ft^3	l	m^3
From:	Multiply by:					
US gallon (gal)	1	0.8327	0.02381	0.1337	3.785	0.0038
UK gallon (gal)	1.201	1	0.02859	0.1605	4.546	0.0045
barrel (bbl)	42.0	34.97	1	5.615	159.0	0.159
cubic foot (ft^3)	7.48	6.229	0.1781	1	28.3	0.0283
litre (l)	0.2642	0.220	0.0063	0.0353	1	0.001
cubic metre (m^3)	264.2	220.0	6.289	35.3147	1000.0	1

Unit prefixes

kilo-	k	10^3	1,000
mega-	M	10^6	1,000,000
giga-	G	10^9	1,000,000,000
tera-	T	10^{12}	1,000,000,000,000
peta-	P	10^{15}	1,000,000,000,000,000

For example:
milliwatt (mW): 1000th of a watt
kilowatt (kW): 1000W
megawatt (MW): 1,000,000W
gigawatt (GW): 1,000,000,000W
terawatt (TW): 1,000,000,000,000W. In 2006 about 16TW of power was used worldwide.

Power and energy

Power is the rate at which energy is produced by a generator or consumed by an appliance.
Unit: the watt (W). 1000 watts is a kilowatt (kW).

The amount of power produced by a generator or consumed by an appliance over a period of time is what we commonly think of as 'energy' on a bill.

Units: the watt-hour (Wh). 1000 watt-hours is a kilowatt-hour (kWh). Watt-hours can be used to describe heat energy as well as electrical energy, but joules (J) are also used for heat. 1Wh = 3600 J. A joule is one watt per second, since there are 3,600 seconds in an hour. 3.6 megajoules (MJ) = 1kWh.

Examples:

• One 80W light bulb on for two hours, or two 80W bulbs on for one hour would consume 2h × 80J/s = 160Wh.
• Three 80W light bulbs on for six hours will consume 3 × 80W × 6h = 1440Wh or 1.44kWh.

Energy efficiency and lighting

For the same amount of luminescence over four hours:

• An 80W incandescent bulb will consume 80W × 4h = 320Wh.
• A low-energy 8W LED will consume 8W × 4h = 32Wh.

From this we can see that an LED lumaire is 320/32 = ten times more efficient than an incandescent bulb for the same amount of luminescence. LEDs will also last many times longer.

Energy efficiency, fans and air-conditioning

For similar benefits fans consume much less electrical energy than air-conditioners for a similar level of comfort.

• A typical fan of 30W on for six hours will consume 30W × 6h = 180Wh.
• A typical air-conditioning unit of 2kW (2000W) on for six hours will consume 2000W × 6h = 12,000Wh.

From this we can see that a fan is 12,000/180 = 67 times more efficient than an air-conditioning unit for the same degree of comfort.

Power generation

- One photovoltaic solar panel producing 80W for two hours, or two panels producing 80W for one hour would produce 2h × 80W = 160Wh.
- Three panels producing 90W for five hours will produce 3 × 90W × 5h = 1350Wh = 1.35kWh.

Index

9 781138 806283